An Atlas of Middle Eastern Affairs

The Middle East is a major focus of world interest. This atlas provides accessible, concisely written entries on the most important current issues in the Middle East, combining maps with their geopolitical background. Offering a clear context for analysis of key concerns, it includes background topics, the position of the Middle East in the world and profiles of the constituent countries.

Features include:

- Clearly and thematically organised sections covering the continuing importance of the Middle East, the background, fundamental concerns, the states and the crucial issues related to the area.
- Original maps integrated into the text, placing international issues and conflicts in their geographical contexts.
- Case studies and detailed analysis of each country, complete with relevant statistics and key facts.
- Coverage of fundamental considerations, such as:
  - water shortage;
  - the petroleum industry;
  - conflicts and boundary issues.
- A comprehensive further reading section, enabling students to cover the topic in more depth.

A valuable introduction to undergraduate students of political science and Middle East studies and designed as a primary teaching aid for courses related to the Middle East in the areas of politics, history, geography, economics and military studies. This book is also an outstanding reference source for libraries and anyone interested in these fields.

Ewan W. Anderson is currently Visiting Professor in the Institute of Arab and Islamic Studies at the University of Exeter, and Emeritus Professor of Geopolitics at the University of Durham.

Liam D. Anderson is an Associate Professor of Political Science at Wright State University, Dayton, Ohio, where he teaches classes on international relations and comparative politics. He specialises in issues of constitutional design, particularly in the context of Iraq and other divided societies.

An Atlas of Middle Eastern Affairs

Ewan W. Anderson and Liam D. Anderson

Cartography by Ian M. Cool

Routledge
Taylor & Francis Group

LONDON AND NEW YORK

First published 2010
by Routledge
2 Park Square, Milton Park, Abingdon, Oxon OX14 4RN

Simultaneously published in the USA and Canada
by Routledge
270 Madison Ave, New York, NY 10016

Routledge is an imprint of the Taylor & Francis Group, an Informa business

© 2010 Ewan W. Anderson and Liam D. Anderson

Typeset in Times New Roman by Saxon Graphics Ltd, Derby
Printed and bound in Great Britain by the MPG Books Group

British Library Cataloguing in Publication Data
A catalogue record for this book is available from the British Library

Library of Congress Cataloging in Publication Data
Anderson, Ewan W.
  An atlas of Middle Eastern affairs / Ewan W. Anderson and Liam D. Anderson ; cartography by Ian Cool.
    p. cm.
  Includes bibliographical references and index.
  ISBN 978-0-415-45514-5 (hardback : alk. paper) -- ISBN 978-0-415-45515-2 (pbk. : alk. paper) -- ISBN
978-0-203-87136-2 (e-book) 1. Middle East--Politics and government--1979- 2. Middle East--Politics and
government--1979---Maps. 3. Geopolitics--Middle East. 4. Geopolitics--Middle East--Maps. I. Anderson, Liam
D. II. Cool, Ian. III. Title.
  DS63.1.A535 2009
  956.05--dc22

                                                  2009008966

ISBN 978-0-415-45514-5 (hbk)
ISBN 978-0-415-45515-2 (pbk)
ISBN 978-0-203-87136-2 (ebk)

To
Conor and Kiernan Anderson
&
Ewan and Isla Cool

May their generation prove wiser and more successful
than those that have gone before.

Contents

Maps

Section D

Section E

Tables

Preface

Located at the crossroads of the world where the three Old World continents meet, the area distinguished today as the Middle East has always been of more than regional significance. It saw the birth of civilisations and was the origin of three great monotheistic world religions. It saw the clash of empires but also the growth of Arab and Islamic domination. Throughout history, it has remained crucial as a focus of routeways and movement but since the Second World War it has captured international attention as a result of two key factors. The first was the establishment of the state of Israel within the Arab core of a region which is overwhelmingly Islamic. The second was the occurrence of the world's major reserves of petroleum. The Middle East can be considered the world's premier flashpoint.

Superimposed upon the *mélange* of issues that afflict the Middle East is the fact that, with a few notable exceptions, the states of the Middle East remain little visited by people from the Western World. The region retains an innate mystique and few are able to make judgements from personal experience. To the many, the mention of the Middle East produces a reflex action. The region represents the major set of problems on the global stage. As a result, there are likely to be as many misperceptions as there are accurate perceptions of the Middle East.

The aim of this *Atlas* is to present an accessible yet manageable coverage, in maps and text, of the affairs of the Middle East. The essential summary is provided by the maps and the text has been used to illuminate the key points in a concise manner. A list of further reading is provided for those who require greater detail. In subject matter, the *Atlas* follows a progression from a consideration of the basic landscape elements and the fundamental common concerns to the states themselves, the effects of contiguous states and the key issues which result from the actions of the states.

Section A puts the Middle East in context. The geographical and historical background of the region, in as much as it provides the stage for the actors, is set out in Section B. The fundamental components: petroleum, water, boundaries and transboundary movements are examined in Section C. Section D provides a synopsis of each of the states and territories together with the groups of contiguous Rimland regions. For each state, the key facts are tabulated, a list of recent events is provided and there is discussion of the major issues faced by the state and its current status on the global stage. For the Rimland regions, the focus is upon their relations with the Middle East. Section E offers a description and analysis of the recent and current key issues in the Middle East.

The main sources of statistics have been *The World Factbook* (Central Intelligence Agency 2008), *The Military Balance* (International Institute for Strategic Studies 2008), the *BP Statistical Review of World Energy* (June 2008) and the following volumes of the Europa Regional Surveys of the World (Routledge): *The Middle East and North Africa 2007, South Asia 2007, Africa South of the Sahara 2007, Eastern Europe, Russia and Central Asia 2007*. We are particularly grateful for access to these definitive volumes. Other sources are noted in the text.

We would place on record our thanks to Joe Whiting who not only suggested the production of the volume but offered insightful support throughout the work. We are also extremely grateful to Rosemary Baillon who managed the overall procedure, organised the volume and typed a large part of the text. Any errors are ours alone.

Ewan Anderson	Liam Anderson	Ian Cool
Exeter, UK	Dayton, USA	Durham, UK

January 2009

Abbreviations and acronyms

AQI	Al Qaeda in Iraq
AQM	Al Qaeda in the *Maghrib*
BC	Before Christ
bcm	billion cubic metres
BP	British Petroleum
BWC	Biological Weapons Convention
CENTCOM	Central Command
CIA	Central Intelligence Agency
CIS	Commonwealth of Independent States
CL	Civil Liberties
CPA	Comprehensive Peace Agreement
cu km/yr	cubic kilometres per year
cu m/y	cubic metres per year
CWC	Chemical Weapons Convention
DOP	Declaration of Principles
E	east
EEBC	Ethiopia–Eritrea Boundary Commission
EEZ	exclusive economic zone
EOKA	National Organisation for Cypriot Struggle
EU	European Union
F	free
FIS	Islamic Salvation Front
FMA	Foreign Military Assistance
GAP	South-east Anatolian Project
GCC	Gulf Cooperation Council
GDP	gross domestic product
GHQ	General Headquarters
HEU	highly enriched uranium
IAEA	International Atomic Energy Authority
ICJ	International Court of Justice
IDF	Israeli Defence Force
IDP	internally displaced persons
IHO	International Hydrographic Organisation
ISCI	Islamic Supreme Council of Iraq
ISI	Inter-Services Intelligence
ITG	Iraqi Transitional Government
JEM	Justice and Equality Movement
KDP	Kurdistan Democratic Party

KDPI	Kurdistan Democratic Party of Iran
KGK	People's Congress of Kurdistan or Kongra-Gel
km	kilometre
km²	square kilometre
KRG	Kurdistan Regional Government
LEU	low-enriched uranium
m	metre
m²	square metre
m³	cubic metre
mcm/y	million cubic metres per year
mill tonnes	million tonnes
MINURSO	UN Mission for the Referendum in Western Sahara
mm	millimetre
N	north
NA	Northern Alliance
NATO	North Atlantic Treaty Organisation
NF	not free
nml	nautical miles (1 nml = 1.852 km)
NPT	Non-proliferation Treaty
NSS	National Security Strategy
OAU	Organisation of African Unity
OPEC	Organisation of Petroleum Exporting Countries
OSCE	Organisation for Security and Cooperation in Europe
PA	Palestinian Authority
pcm	per cubic metre
PDRY	People's Democratic Republic of Yemen
PF	partly free
PKK	Kurdish Workers' Party
PLC	Palestinian Legislative Council
PLO	Palestine Liberation Organisation
Polisario Front	Popular Front for the Liberation of the Saguia el Hamra and Rio de Oro
PR	political rights
PUK	Patriotic Union of Kurdistan
R&D	research and development
R/P	reserves to production
RUSI	Royal United Services Institute
S	south
SADR	Sahrawi Arab Democratic Republic
SCIC	Supreme Council of Islamic Courts
SCIRI	Supreme Council for the Islamic Revolution in Iraq
SLA	South Lebanon Army
SLM	Sudanese Liberation Movement
SUMED	Suez–Mediterranean oil pipeline
TAL	Transitional Administrative Law
TFG	Transitional Federal Government
TNA	Transitional National Assembly
TRNC	Turkish Republic of Northern Cyprus
TVPA	Trafficking Victims Protection Act
UAE	United Arab Emirates

UAR	United Arab Republic
UK	United Kingdom
UN	United Nations
UNCLOS	UN Conference on the Law of the Sea
UNDOF	UN Disengagement Observer Force
UNEF	UN Emergency Force
UNESCO	UN Educational, Scientific and Cultural Organisation
UNHCR	UN High Commission for Refugees
UNIFCYP	UN Force in Cyprus
UNIFIL	UN Interim Force in Lebanon
UNMEE	UN Mission in Ethiopia and Eritrea
UNRWA	UN Relief and Works Agency
UNSC	UN Security Council
UNSCOM	UN Special Commission
UNTSO	UN Truce Supervision Organisation
US	United States (adj.)
USA	United States of America
USS	United States Ship
VX	chemical nerve agent
W	west
WMD	weapons of mass destruction
YAR	Yemen Arab Republic
°	degrees
°C	degrees Centigrade
%	per cent
$	US dollars
×	times
–	not available / not applicable / not listed / none
9/11	11 September 2001

Section A

The Middle East in context

Throughout history the Middle East has been an important focus of human affairs. Soon after the last glacial period (1200–1500 BC) within the region, the settled occupation of the land with the cultivation of crops occurred. This change from nomadic hunter gathering is recognised as the first real stage in civilisation. With the increasing organisation necessary to establish large-scale cultivation and the accumulation of surplus wealth, the first cities and the first recognisable political units appeared in the basins of the Tigris–Euphrates and the Nile.

Its location at the centre of the World Island (Map 1) allowed the transfer of ideas from all directions. The Middle East was the global communications hub which attracted not only prosperity but also conflict. It saw conflict between empires and states but also the birth of great world religions. Through the blossoming of Arab arts and science, it linked the world of the Ancients with that of the Renaissance. The Middle East was highly significant in both world wars and, subsequently, its oil wealth has made it a key political flashpoint. Since the Second World War, the establishment of Israel in a totally Arab and Islamic region has ensured the Middle East a continuing high global profile and a close association with conflict.

Despite the fact that much of its history is lost in the mists of time, the actual term 'Middle East' is a relatively recent acquisition. Since the fifteenth century and the Great Age of Discovery it had become customary to distinguish between the Near East and the Far East. The former comprised the eastern Mediterranean with its adjacent lands while the Far East was everything east of India. The Indian subcontinent itself occupied something of an anomalous position. In terms of trade and therefore military strategy, it was effectively the centre of the British Empire but, other than its most northerly and westerly approaches, was considered a separate entity, located in the East but not the Near East or the Far East. Thus, almost by default the area located in the middle between the Indian subcontinent and the Near East was characterised as the Middle East. Therefore, historically and geographically there is some logic in the designation of the Middle East or what is known as the *Mashreq*.

Although the term Middle East was used in the British India Office during the middle part of the nineteenth century, its introduction to a wider audience is normally credited to Alfred Mahan, an American naval officer, geopolitician and historian. The core of the Middle East as seen by Mahan was the Gulf. Later, his terminology was variously interpreted as the 'Middle' in an east–west sense but also possibly in a north–south sense between the British to the south and Russians to the north. Thus, from about 1900, the term 'Middle East' was no longer restricted to the corridors of power in the British government but came to have a wider acceptance.

However, it was during the period of global warfare that the term achieved general acknowledgement and permanence. During the First World War, the operational area of the Mesopotamia Expeditionary Force was distinguished as the 'Middle East' while that of the Egyptian Expeditionary Force was characterised as the 'Near East'. Geographical and strategic or military definitions roughly coincided. Between the wars, the Royal Air Force amalgamated its Middle Eastern

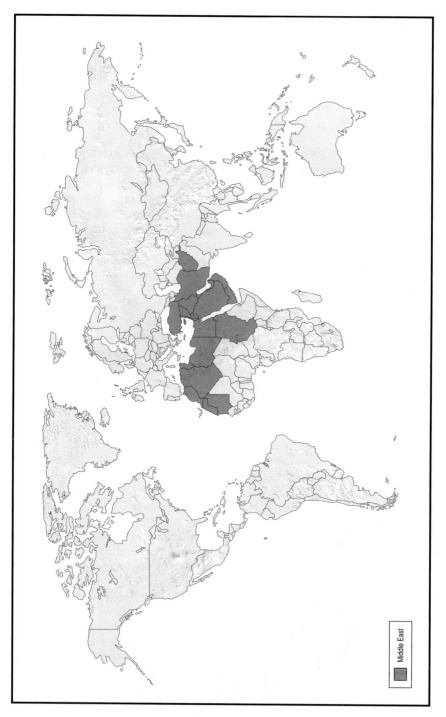

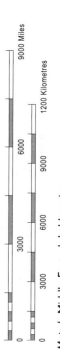

Map 1 Middle East: global location

Command based in Iraq, with its Near Eastern Command located in Egypt. The new Command retained the term 'Middle East'. This precedent was followed by the Army, which at the beginning of the Second World War, largely for the defence of the Suez Canal, established its General Headquarters (GHQ) Middle East in Cairo. Thus a swathe of land extending from Iran to Tripolotania was created as a military province and given the name 'Middle East'. The importance of this region, especially in political and economic terms, resulted in the appointment of a Minister of State and the development of an economic organisation known as the Middle East Supply Centre. Originally the Supply Centre was British but later it became Anglo-American and thereby the term 'Middle East' became generally recognised not only in the Eurocentric world but also ironically in the home of its originator.

Following the Second World War, the Middle East became the standard term of reference for a wide variety of activities: military, political and economic. At the same time, the term 'Near East' gradually faded from common usage. Throughout the Cold War, the Middle East remained a focus of attention for both the East and the West. The United States of America (USA) and the Soviet Union developed close relationships with various states in the region and some, such as Egypt, changed sides. As a result, the term 'Middle East' came into use worldwide, including within the region itself. However, while the term may be generally accepted, there has never been agreement on the exact boundaries of the Middle East. The United States (US) State Department has compromised by using the term 'Near and Middle East'. In the US military, the region is divided between three Commands: the European, the Middle Eastern and the African.

With regard to boundaries, the chief problem would seem to lie in the west where Arab and Islamic countries are located far from any geographical area which could be considered both 'eastern' and 'middle'. Indeed, Mauritania extends to almost the western extremity of Africa. Such considerations have given rise to the term 'Middle East and North Africa'. However, even with that, geography needs to be stretched to include Mauritania. To the east, Iran, as a Gulf state offers a generally accepted limit to the Middle East. However, Afghanistan, also a Muslim state, is then left in isolation. It is not accepted as part of Central Asia and it is clearly not part of the Indian subcontinent. It was indeed established as a buffer state between the British and Russian influence, ostensibly to keep the Russians out of the Indian subcontinent. It might be suggested that in Mahan's terms, Afghanistan is classically in the middle and the east.

In the eastern part of North Africa, Ethiopia with its powerful Christian heritage offers a clear restriction on Middle Eastern influences. With the recognition of the predominantly Muslim state of Eritrea, this limit is now less clear-cut. However, the Horn of Africa is a generally recognised region. Other Middle Eastern boundary issues concern Turkey, a small but highly significant part of which is within Europe, and Sudan, which clearly extends southwards into the equatorial region of Central Africa. Given a range of factors from Islam and the predominance of Arabic to lifestyle, together with political, economic and social affiliation, for this *Atlas* the Middle East will be used as a term to include North Africa and to extend from Mauritania in the west to Afghanistan in the east and from Turkey in the north to Sudan in the south (Map 2). This delimitation includes all the states generally accepted as Middle Eastern together with those of the *Maghrib* and its westward extension into Western Sahara and Mauritania. In the east it includes Afghanistan, at the present time closely linked in the public mind with Iran, Iraq, conflict and terrorism.

Thus defined, the Middle East extends from approximately 41° N to 3° N and from 17° W to 75° E. That is a latitudinal spread of approximately 38° and a longitudinal extent of 92°. It includes twenty-three states together with Palestine, a recognised independent territory, and Western Sahara, the status of which has still to be agreed. The Middle East lies between the Atlantic and Indian oceans and subsumes at least part of five seas: the Red Sea, the Gulf, the Caspian Sea, the Black Sea and the Mediterranean Sea. It lies centrally between the three Old

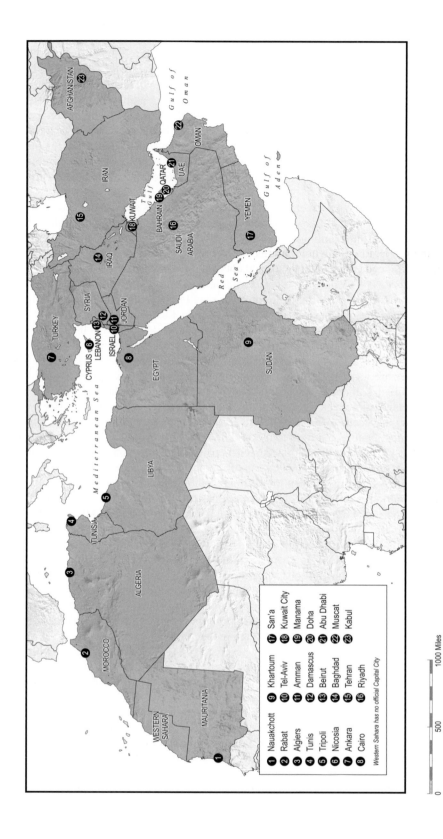

1	Nauakchott	9	Khartoum	17	San'a
2	Rabat	10	Tel-Aviv	18	Kuwait City
3	Algiers	11	Amman	19	Manama
4	Tunis	12	Damascus	20	Doha
5	Tripoli	13	Beirut	21	Abu Dhabi
6	Nicosia	14	Baghdad	22	Muscat
7	Ankara	15	Tehran	23	Kabul
8	Cairo	16	Riyadh		

Western Sahara has no official Capital City

0   500   1000 Miles

0 500 1000 1500 Kilometres

Map 2 Middle East: states and capital cities

World continents of Europe, Asia and Africa, defining in part the continental boundaries by the Red Sea in its northward extension through the Gulf of Aqaba and the Turkish Straits.

The states themselves vary in size from those such as Sudan, Algeria and Saudi Arabia considered large in global terms to micro-states such as Qatar, Bahrain and Kuwait. The states vary enormously in wealth from the oil-rich countries such as the United Arab Emirates (UAE), Qatar and Kuwait to the very poor including Sudan, Western Sahara and Mauritania. All bar Israel, Cyprus and Lebanon are almost exclusively Muslim and all except Lebanon and Cyprus include areas of marked aridity. The Middle East is characterised by unity on a large scale but also by diversity, frequently on a smaller scale. In so many ways, the Middle East is a place of extreme contrasts. The ancient *suq* is set against modern high-rise buildings. A *dhow* slowly makes its way between the tower blocks of Dubai Creek. A Bedouin tent together with its long-term inhabitants appears in the grounds of a multinational-run hotel. Parched desert abuts against irrigated luxuriance.

In viewing the region as a whole, there are a number of outstanding characteristics. Using a polar projection, it can be seen that the region abuts on to all the major global cultures other than those of North and South America (Map 3). It shares boundaries with Europe, Russia through the Commonwealth of Independent States (CIS), Asia-Pacific through China, the Indian subcontinent and Africa. The Middle East is the global transport node and also the key international choke point. Ancient route ways, now replaced by all-weather roads, crisscross the region. Since the opening of the Suez Canal in 1869, shipping has converged on the region and its many choke points: the Straits of Gibraltar, Hormuz and Bab al Mandab and the Turkish Straits. In many ways, the major choke point remains the Suez Canal itself. For air travel to Asia-Pacific, Central Asia and much of Africa, the Middle East is a stopover or plane change point. There is a sense that transport passes through the region but the pilgrimage to Mecca, the *Hajj*, and the petroleum industry are reminders that the Middle East is also a key destination. The most recent transport development has concerned pipelines from the oilfields to the coastal loading point as well as to other parts of the region to avoid choke points or shorten the sea passage.

The key factor most commonly associated with the Middle East is petroleum. Five states in the Gulf region have between them more than two-thirds of all the world's reserves. Oil is the most traded commodity and the complex infrastructure of the industry reaches every country. Petroleum is truly a global industry and, following the demise of coal, it is the key, and in many cases the only, fuel mineral.

The Middle East is perhaps the only world region which can be classed in the true sense as religious. The Middle East is located at the core of Islam (Map 4). The daily calls to prayer dominate every settlement and prayer mats are produced whatever the occasion or environment. Nonetheless, Islam is divided between Sunni and Shi'a, with a number of smaller subdivisions. It must also be remembered that in many of the countries there are significant non-Muslim minorities. Cyprus is predominantly Christian while Israel is Judaic and Lebanon has a high proportion of Christians. Egypt, despite being a core Muslim country, has a significant Christian element. The Middle East is, most importantly, the birthplace of three great world monotheistic religions, while a fourth evolved nearby and has exerted a strong influence.

Arabic is the predominant language throughout the Middle East but, unlike Islam, is not all-pervasive. Of the three regional super-states, only Egypt is Arab. Both Turkey and Iran are foci of their own language groups. There are other minority languages but essentially the Middle East has three key languages. It remains an astonishing fact that, although dialects vary some-what, an Arab from Oman in the east can understand an Arab from Mauritania in the west. In no other world region do such a large number of states share a common mother tongue.

A further key characteristic is the political significance of the region. During the Cold War it lay between East and West with competition between the Soviet Union and the USA. Oman had

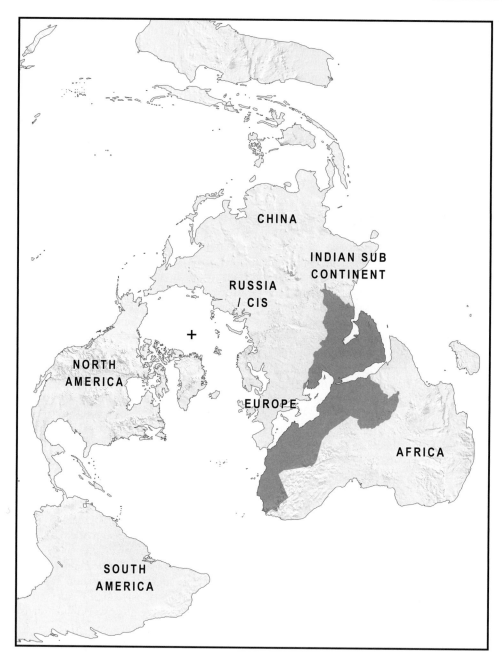

Map 3 Polar projection of the Middle East

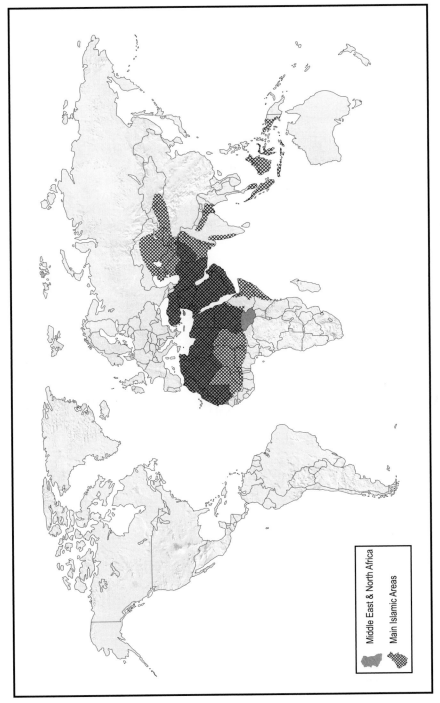

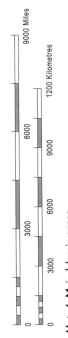

Middle East & North Africa

Main Islamic Areas

0   3000   6000   9000   1200 Kilometres

0   3000   6000   9000 Miles

Map 4 Main Islamic areas

Western bases such as Masirah Island while down the coast off what was the People's Democratic Republic of Yemen, Socotra was a base for the Soviet Navy. However, despite the number of bases and the flow of arms into the region, only one state, Turkey, was a member of a military alliance, the North Atlantic Treaty Organisation (NATO). At the end of the Cold War, NATO remained in being as the Warsaw Pact disintegrated. In justifying its existence, NATO cited the possible threat of Islamic Fundamentalism from the Middle East. Almost by wish fulfilment, violence within the region increased. However, it reached a gory peak following the US/United Kingdom (UK) invasion of Iraq. Numerous groups, characterised by the State Department as terrorists, joined those already in existence in the region. As a result, although on any real scale it is only a recent phenomenon, the region has become associated with terrorism and its generation. The other obvious source of violence followed the foundation of the state of Israel. Israel is in the Middle East but not of the Middle East in that its living standards and aspirations are far higher than those of any other country. If the other states are Third World or at best Second World, Israel is essentially First World and it seeks closer links with the USA and Europe rather than with its neighbours. The forcible emplacement of a Jewish state in the middle of an area that is wholly Arab and Muslim was always likely to lead to problems and violence. Over the years attitudes have hardened, exacerbated by the stance of the USA which has offered unequivocal support for Israel. Therefore, in the public mind, the Middle East tends to be associated with violence and conflict whether or not these are generated within the region.

Section B
The Middle Eastern background
Geographical and historical

States and territories of the Middle East can be envisaged as actors in a continuing drama. Each has a unique character and each exhibits different abilities. For example, the power to influence other states may depend upon location, history, area of territory, size of population, economic strength, military strength, external support or the contribution of significant individuals. The characteristics of each state, together with those of neighbouring groups of states are discussed in Section D.

While the drama may, and frequently does, affect the whole world, the essential stage on which the action takes place is the landscape of the Middle East. This landscape, with its physical and human components, is the resultant of geography and history. Therefore, if Middle Eastern affairs are to be identified and interpreted, a basic appreciation of these two subjects is required.

Each state brings its own geography and history to the play but the drama can only be appreciated in context against the broad sweep of Middle Eastern geography and history. Key elements of the total physical landscape, such as aridity, have been influential throughout the history of the region and remain crucial in current Middle Eastern affairs. Islam is a dominant component of the human landscape and the influence of this religion pervades all areas of the Middle East.

It is with such thoughts in mind that the background will be described. Aspects of the background are discussed only if they are considered relevant to an understanding of Middle Eastern affairs. However, it is realised that, given time, even the most seemingly insignificant element might be, for some activity, vital. Over geological time, no variation is too small to be discounted, but if the background is to relate directly to the other sections of the *Atlas* there needs to be selection and, given the scale of the region, there must also be generalisation. As an example, vegetation is clearly an important landscape feature but it varies as a result of soil, climate, geomorphology and a number of biotic factors, particularly human influence. Since soil variation itself is too complex to map on such a regional scale, it can be seen that only the most generalised picture of vegetation can be provided.

Present-day geography sets the stage on which Middle Eastern events occur but it is by no means passive. The geography of the Middle East is, in many ways, the geography of extremes and, as such, makes contributions to the script by imposing constraints and limitations as well as by offering opportunities. The current landscape is, furthermore, only a snapshot of a continuous process of evolution whether one considers, for example, landforms or settlement patterns. Relatively recent changes, particularly those involving the human landscape, are considered history. Middle Eastern affairs may result not only from the current combination of factors but they may have a significant historical legacy. For convenience and because they represent different viewpoints, one focusing on time and the other on place, the historical and geographical perspectives will be considered separately although it must always be borne in mind that they are intimately entwined.

Geographical background

The present landscape of the Middle East, more than that of any other world region, illustrates the sharpest contrast between the works of nature and the works of man. Landscapes in the temperate zone are basically the product of natural variables and human factors. Mountainous areas may have been little affected by human activity while urban areas may seem little influenced by nature, but in general there is a clear interplay between the two. In the Middle East, there are vast areas, some mountainous but mainly comprising desert, in which any effect of human life is difficult to detect. In contrast, the valleys of the Nile, the Tigris and the Euphrates have been moulded by human action over millennia. Given this contrast, it is appropriate to separate considerations of physical geography from those of human geography.

Physical geography

As indicated, physical geography is important in any consideration of Middle Eastern affairs in that it places obvious constraints on human activity. To name but a few aspects, agriculture, movement and indeed permanent settlement are all severely limited in the Middle East. However, the geological conditions for the formation of petroleum and natural gas, the river valleys with their possibilities for agriculture and settlement and the deep aquifers as a source of water all offer opportunities for human development. Thus, aspects of both the underlying structure and the surface form are of significance.

Relief and structure

Given the fact that so much of the landscape is unencumbered by human artefacts or even vegetation, the relationship between structure and form is, in most parts of the Middle East, very apparent. However, underlying this clarity and seeming simplicity is complexity of scale and process. The foundations laid during tectonic history are on a macro-scale. The geological sequence, a product of emergence and submergence together with drier and wetter periods, varies over time and in location throughout the region and the differences are essentially on a meso-scale. The moulding of the present landscape has been by geomorphological processes, many of which operate on a micro-scale.

Interpretation is further hindered by the fact that there is variation in the current state of knowledge. There has been elaboration of the theory of plate tectonics and particularly the locations of the edges of the plates but with regard to geology the level of knowledge varies widely across the Middle East. In areas in which there has been drilling for potential hydrocarbons or occasionally water, there is abundant information about the underlying stratigraphy. In areas lacking these interests, geological information is sparse. Geomorphology depends essentially upon meticulous fieldwork and detailed monitoring, both of which are problematical in many of the harsh environments of the Middle East. It remains true that there are fewer textbooks on arid zone geomorphology than there are annual journals on glaciology.

Tectonics, geology and geomorphology are all closely interlinked. For example, tectonic movements resulting in emergence and submergence control the geological succession and through lateral movements they produce fold mountains and depressions. The changing face of the landscape influences geomorphological processes which are further affected by geology as a result, for example, of permeability. The basis for the large-scale movements is plate tectonics. The surface of the Earth comprises a series of plates, the movement of which produces changes at the plate edges. The movement itself may be accompanied by faulting, earthquakes and volcanic activity and these facilitate the identification of the plate edges. There are three categories of plate edge effects. Plates may move apart in what is termed "spreading", as for example in the case of the central trough of the Red Sea. Alternatively, plates may move together with one plate being compressed against another. One plate is forced below the other and its edge is then consumed in what is termed a "subduction zone". The Taurus and Zagros mountains indicate a subduction zone. Third, plates may move laterally with respect to each other producing transform faults such as that which dictates the basin form of the Jordan Valley.

For the Middle East, a simplified version of events is that there were two major plates, the Afro-Arabian to the south and the Eurasian to the north. Between them a major depression in the Earth's crust was occupied by the Tethys Sea in which, from the erosion of the two plates, vast amounts of sedimentation occurred. Later the Afro-Arabian plate moved northwards thrusting up massive fold mountain ranges of which the mountains in the north of the region are but remnants. As movement occurred, the African plate was split into three separate plates: the African, the Arabian and the Somalian plates, separated by the Great Rift Valley system which runs from the Dead Sea through the Red Sea into the Rift Valleys of Ethiopia and Kenya. Also, the northern edges of the plate shattered producing the Aegean and Turkish plates. The overall result is that the physical landscape of the Middle East can be basically divided into two forms: the continental platforms, representing the plates, and the mountain orogenic belts at the edges of the plates. The platforms comprise stable blocks while the orogenic belts are liable to instability.

Other than the fact that the oldest rocks occur only on the continental platforms, there is little basic difference geologically between the two forms. For obvious reasons, outcrops tend to be far more extensive on the undisturbed platforms than in the orogenic belts. With human activity in mind, a key distinction is between the older sedimentary rocks which tend to occur towards the north and south of the region and the younger sedimentaries which occur through the centre. The former, for example Nubian Sandstone, are generally of continental origin whereas the later are maritime. This position can be related to the tectonic model with marine sediments, particularly limestones and marls, being laid in the Tethys Sea. These calcareous rocks, which occur in the northern part of North Africa and in a belt from the Levant down either side of the Gulf, are particularly vital as it is in them that the petroleum reserves are concentrated.

With regard to geomorphology, a key point is the extent to which the present landforms represent current processes rather than those which obtained during past climates. Over large areas of the region, extensive drainage networks are etched into the landscape even though today they rarely carry any surface water. The issue is whether these channels, called wadis, some of which are massive gorges, could have been produced under present conditions. The prevalent geomorphic processes are aeolian but it is doubtful whether wind action alone can produce landforms on the scale found in many parts of the Middle East. Therefore, the dominant processes may still be fluvial despite the rare occurrence of precipitation. With the current state of knowledge, it can be summarised that in the Middle East the key determinants of landforms are present-day aridity or hyperaridity and occasional intense water action combined with the effects of previous pluvial periods. The distinctive assemblages of landforms in the Middle East are those comprising:

i slopes with their covering of deposits ranging in size from boulders to fine gravel and dust together with the attendant mass movement features such as landslides;
ii wadi networks;
iii dune fields or sand seas (ergs) with their variety of aerodynamically shaped depositional landforms.

The resultant of the tectonics, geology and geomorphology is the relief (Map 5). The two dominant features are the platforms, vast, gently sloping plains, much of the area less than 500 m above sea level. The African platform provides the physical landscape for all of the North African states included in the Middle East other than the northern parts of Morocco, Algeria and Tunisia, where the orogenic belt occurs. The Arabian platform includes the whole of the peninsula and it extends northwards to underlie much of Iraq, south-eastern Turkey and the whole of Israel, Lebanon and Syria. In each case there are higher areas. In Africa there are vast dissected relict plateaux, the most prominent of which are the Hoggar Massif, the Tibesti Plateau and Jebel Marra, all of which exceed 3,000 m above sea level. The Red Sea Hills are similar in origin but are distinguished by steep fault-line scarps along the Red Sea littoral. These vast plains are predominantly covered in lag gravels, interspersed with bare rock areas, sand seas or ergs and occasional areas of salt marsh or sebkha. In the Arabian Peninsula, the Red Sea is paralleled on its eastern side by uplands which culminate to the south in the Asir Mountains which reach well over 3,000 m above sea level. As with the Red Sea Hills to the west of the Red Sea, these uplands are distinguished by sharp fault-line scarps above the eastern littoral of the Red Sea.

In Africa, the great ergs include the Rebiana sand sea which extends from western Egypt and is a major feature of the Libyan Desert. In Algeria there are the two parallel sand seas, the Grand Erg Occidental and the Grand Erg Oriental which extends into Tunisia. The Erg Iguidi occupies a large part of northern Mauritania and a significant area in western Algeria. In the Arabian Peninsula, the largest sand sea is the Rub al-Khali or "Empty Quarter" which covers an area of 560,000 km². To the north is the Nafud another vast area of sand and to the south-east the Wahiba Sand Sea of Oman with an area of approximately 20,000 km². The sand seas are characterised by their own landforms: barchans (crescentic dunes), seif (linear dunes), star and compound dunes which may reach heights of over 300 m. Major saline landscapes occur in the plateau of the Shotts in southern Tunisia and the Qattara depression in western Egypt, a vast deflation (wind-eroded) hollow which reaches 133 m below sea level.

In western Jordan, Israel and Lebanon, there are upland areas but the relief is dominated by the system of faults which constitute the major north–south trough running from the Jordan Valley through the Dead Sea to the Gulf of Aqaba. The enclosing highlands may reach over 2,000 m above sea level in places and the floor of the Dead Sea basin is more than 300 m below sea level.

The orogenic belts to the north of the great platforms comprise young fold mountains, which dominate the landscape in northern Morocco, Algeria, Tunisia, eastern Iraq and a large part of Turkey and Iran. This belt of highland which extends basically in an east-west direction is part of the Alpine system which includes the mountains of southern Europe and the Himalayas, an outlier of which is the Hindu Kush, the mountainous core of Afghanistan. In the *Maghrib*, there are several west-east trending ranges which, in the High Atlas of Morocco, reach a height of 4,162 m. There are three discernable Atlas ranges in Morocco and two in Algeria which converge to form the Aures Mountains, extending into Tunisia.

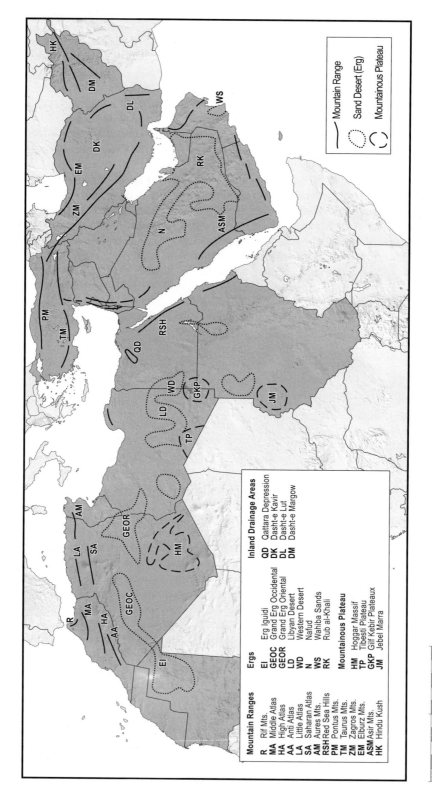

Mountain Ranges

R	Rif Mts.
MA	Middle Atlas
HA	High Atlas
AA	Anti Atlas
LA	Little Atlas
SA	Saharan Atlas
AM	Aures Mts.
RSH	Red Sea Hills
PM	Pontus Mts.
TM	Taurus Mts.
ZM	Zagros Mts.
EM	Elburz Mts.
ASM	Asir Mts.
HK	Hindu Kush

Ergs

EI	Erg Iguidi
GEOC	Grand Erg Occidental
GEOR	Grand Erg Oriental
LD	Libyan Desert
WD	Western Desert
N	Nafud
WS	Wahiba Sands
RK	Rub al-Khali

Mountainous Plateau

HM	Hoggar Massif
TP	Tibesti Plateau
GKP	Gilf Kebir Plateaux
JM	Jebel Marra

Inland Drainage Areas

QD	Qattara Depression
DK	Dasht-e Kavir
DL	Dasht-e Lut
DM	Dasht-e Margow

Legend:

⎯ Mountain Range
⋯ Sand Desert (Erg)
⌒ Mountainous Plateau

0   500   1000 Miles
0   500   1000   1500 Kilometres

Map 5 Major landscape features

Between the Alpine ranges are high plateaux, the best example being the Anatolian Plateau of Turkey. To the north of the plateau are the Pontus Mountains and to the south the Taurus Mountains, each comprising several ranges. The two systems converge in eastern Turkey in the Mount Ararat Massif which reaches a height of 5,165 m. In Iran, the ranges diverge, the Zagros Mountains extending south-eastwards to provide a backing for the Gulf. The Elburz Mountains enclose the southern end of the Caspian Sea and rise to a height of 5,610 m. Along the eastern boundary of Iran several small ranges reach heights of over 3,000 m. Between the Elburz Mountains and the Zagros Mountains is a large plateau area enclosing two main basins of internal drainage. To the north is the Dasht-e Kavir a major salt desert, still developing under present-day conditions, and to the south is the Dasht-e Lut, predominantly a rock desert but with extensive areas of sand dunes. Eastwards, Afghanistan is dominated by the ranges of the Hindu Kush and its extensions which fan out across the country south-westwards from the Pamir Knot. The most extensive plain to the south is the Dasht-e Margow, an arid desert area of internal drainage which focuses upon the Hamuun-e Saberi, an extensive area of saline landscape immediately to the west in Iran.

Climate

The core of the Middle East lies between latitudes 30° and 20° N, a belt characterised globally by aridity. The region as a whole is located between the warm temperate climates, including the Mediterranean climate (wet winter and summer drought), and the equatorial climates (hot and wet all the year). The Black Sea coast of Turkey has a Mediterranean climate while the southern tip of Sudan is essentially equatorial. The causes of aridity in the sub-tropical zone, between these two climatic belts, are related to the general circulation of the atmosphere. There is convergence at higher levels which results in subsiding air and divergence at the surface. The subsiding air warms and creates stable conditions inimical to convection and the formation of rainfall. Related to this is the occurrence of a high pressure zone near the area of latitude 30° N. It is also thought that the aridity is related to the activity of the jet stream (a meandering belt of fast moving air) in the upper atmosphere.

Whatever the cause, the result is that limitations are placed upon human activity predominantly by aridity but also by temperature. The two key climatic types are recognised as steppe or semi-arid and desert or arid. The former can be classified as cool and dry with a mean annual temperature below 18° C and the latter as hot and dry with a mean annual temperature above 18° C. However, there are variations throughout the region depending principally upon: latitude, distance from the sea or continentality, altitude and specific meteorological systems such as depressions or upper air controls. Climate and meteorology comprise a range of variables but the complexities of the pressure systems are of direct relevance to human activity most obviously through temperature and rainfall.

The main features of temperature in the Middle East are the high temperatures of summer and the wide range of temperature both diurnal and annual. Mean annual temperature range varies from 5° C in southern Sudan to over 25° C in parts of Syria, Turkey and Iran. During the day, predominantly cloudless skies allow intense solar heating of the land surface but at night there is little check on heat radiation and therefore everywhere except near maritime influences temperatures fall considerably. Temperatures are reduced with altitude, most particularly in winter. July is normally the hottest month and the following mean monthly temperatures for that month reflect the intensity of heating and also the effect of maritime influences: Algiers 24°; Cairo 28°; Khartoum 34°; Tehran 30°; Abadan 35°; Damascus 27°; Baghdad 33°. January is normally the coldest month and mean temperatures for that month are: Algiers 12°; Tunis 10°; Cairo 13°; Khartoum 23°; Tehran 2°; Abadan 12°; Damascus 7°; Baghdad 10°. Extremely low temperatures occur in much of Iran, as indicated by the existence of one small permanent glacier in the Zagros Mountains. Erzerum has a January mean temperature of -11° C, an average night minimum of -27° C and an absolute minimum temperature of -40° C. Only the lower areas of the southern Arabian Peninsula and all of Sudan can be guaranteed to be entirely free from snowfall.

With the exception of southern Sudan, which is affected by equatorial influences, the map of rainfall bears a relatively close resemblance to that of altitude (Map 6). Apart from the few limited areas of intensive irrigation in the river valleys, it also bears a passing resemblance to the map of population (Map 10). For most of the Middle East, rainfall, such as it is, occurs during the winter half-year. Together with the low totals, rainfall is characterised by its variability from place to place and particularly from time to time and also by its intensity. The occurrence of rain can be considered to be extremely capricious. The Levant coast receives over 700 mm annually, all in the winter six months but even the wettest months have only between fourteen and eighteen rain days. However, as much as 25 mm may fall in one hour. Damascus has an annual average rainfall of 240 mm but has recorded 100 mm in a single morning. The variability can be illustrated by Cairo which, with an annual average of 22 mm, has experienced as little as 1.5 mm and as much as 64 mm in a year.

The compelling feature of the rainfall distribution map is the fact that between approximately latitudes 30° N and 18° N annual totals are less than 100 mm. In Africa, the Atlas Mountains, the coastlands of Libya and southern Sudan receive more. In the whole of the Arabian Peninsula only the mountains of Asir, their northward extension along the Red Sea coast and the Jebel Akhdar in Oman exceed an annual figure of 100 mm.

The key factor is the relationship between rainfall and agriculture, although clearly the correlation is not as close as might be expected as a result of irrigation. Neglecting the effect of altitude on agriculture, 400 mm might conveniently be taken as the limit of crop production which in areas near the limit comprises dry farming. Areas with a rainfall of over 400 mm occur essentially around the edges of the region: throughout the Mediterranean coastlands and higher areas of Morocco, Algeria and Tunisia; on the Jebel Akhdar in Libya; in most of Cyprus; along the coastland of the Levant; throughout most of Turkey except in certain interior basins; along the Caspian coastline of Iran together with most of north-western Iran and the main ranges of the Zagros and Elburz Mountains; in the Asir Mountains of Yemen; throughout the southern of Sudan; and throughout the mountainous areas of Afghanistan. Iran and Afghanistan have particularly striking contrasts, in both cases arid desert areas abut on to mountain ranges with moderate to high rainfall. Relatively heavy rainfall, in excess of 750 mm, is found along the Mediterranean coastlines of Morocco, Algeria and Tunisia, the Levant and Turkey together with the Black Sea coastline of Turkey, the Caspian Sea coastline of Iran and the higher areas of the Zagros Mountains in north-western Iran. With the sole exception of Sudan, the southern part of which is in a separate climatic belt, rainfall is closely related to relief throughout the Middle East and therefore areas suitable for high-grade agriculture are extremely limited. The water supply in these mountainous areas is supplemented by snow-melt. In the higher areas, snow may be in evidence for as much as six months of the year while the plateau areas of Turkey and Iran together with large parts of Cyprus and Lebanon also benefit from the moisture input from snowfall.

In considering the significance of rainfall for human activity in the Middle East, two other factors need to be taken into consideration. In all but the wettest areas of the Middle East, the effectiveness of rainfall is greatly constrained by temperature. If rates of evaporation and evapotranspiration (evaporation combined with transpiration from plants) are high, the influence of rainfall upon plant growth and for agriculture generally will be severely limited. Statistics are particularly scarce and evapotranspiration is notoriously difficult to measure.

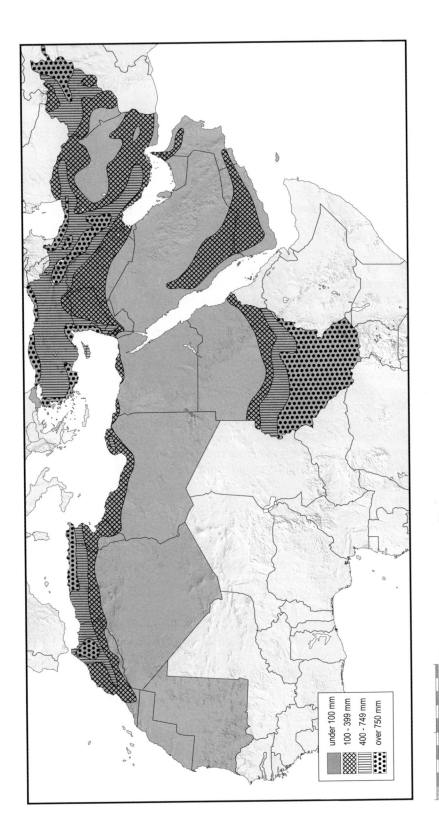

Map 6 Mean annual precipitation

under 100 mm
100 - 399 mm
400 - 749 mm
over 750 mm

0   500   1000 Miles

0   500   1000   1500 Kilometres

Nevertheless, as far as can be ascertained, throughout the drier parts, those with less than 100 mm, mean annual potential evapotranspiration exceeds 1,100 mm. Therefore, while the occasional heavy downpour will stimulate growth, there will normally be a soil moisture deficit. The other factor which affects agriculture is irrigation from surface or ground water.

Hydrology

In the main arid belt of the Middle East, there are only two permanent river systems of any size, those of the Nile and the Tigris–Euphrates. There is surface flow in the case of the Jordan and a few rivers in Lebanon, south-western Saudi Arabia and south-eastern Iran but these are all on a small scale (Map 7). Irrigation from rivers is primarily found in the valleys of the two major systems and in the mountainous areas along the north of the region.

In all the upland areas, the water supply is likely to be supplemented from surficial or at least shallow aquifers. In the drier areas, there may be similar sources towards the mountain fringe but these are likely to be greatly affected by evapotranspiration.

The other major source of supply is from the deep aquifers (Map 8). These contain essentially fossil water, mostly stored from previous pluvial periods, and therefore non-renewable. That these sources can be highly significant is demonstrated by the Great Man-made River from Kufra and Tazerbo both located deep in the Nubian sandstone of the Kufra aquifer. However, relatively little is known about deep aquifers in that, unless through petroleum exploration or scientific necessity, there are no drilling records to indicate the boundaries. Furthermore, apart from the cost of exploitation, some water at least from the deep aquifers is likely to be saline and require desalination. A new school of thought, backed by evidence from many parts of the world, does indicate that, largely through fractured rock and major fault-line aquifers, there may be more connections than originally thought between surface flow and deep aquifers. Of all the factors in the physical landscape, it is water which exercises the most profound influence upon Middle Eastern affairs but, for all aspects of the subject other than the flow of the major river systems, data are very limited.

Soil and vegetation

Since they provide only the surficial covering, the direct effect of soils upon the landscape, other than in extreme cases such as saline soils, is relatively modest. However, the indirect effects through the development of vegetation and the agricultural activities of man are extremely significant. Soils result from the interrelationship between climate and geological parent material, modified by geomorphology, particularly slopes, and a number of organic factors. The broad categories of soil are determined principally by climate and particularly the availability of water. The chief modifiers of this broad pattern are the occurrence of limestone, the effects of salinity, inputs of humus related to vegetation and the effects of human activity. Furthermore, soils develop over time and the rate of formation is slow, particularly so in areas where there is a lack of water.

Given these determinants there is a wide variety of soil types within the wetter areas of the Middle East, influenced locally by such factors as the rainfall total, the occurrence of calcareous rocks, the vegetation covering and agriculture. In the steppe areas, the same modifiers obtain but rainfall is considerably lower, vegetation is sparser and agriculture is far more limited. However, while the lower levels of moisture mean that rock weathering will be considerably slower than in areas of heavier precipitation, there will be considerably less leaching of soil nutrients. Throughout the desert areas of the Middle East, soils are poorly developed and in areas totally

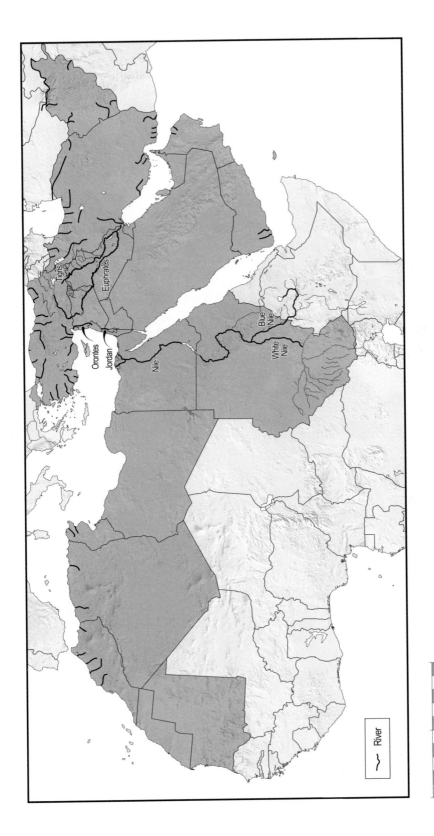

Map 7 Permanent surface flow

River

	1000 Miles
0	500	1000	1500 Kilometres
0	500	1000

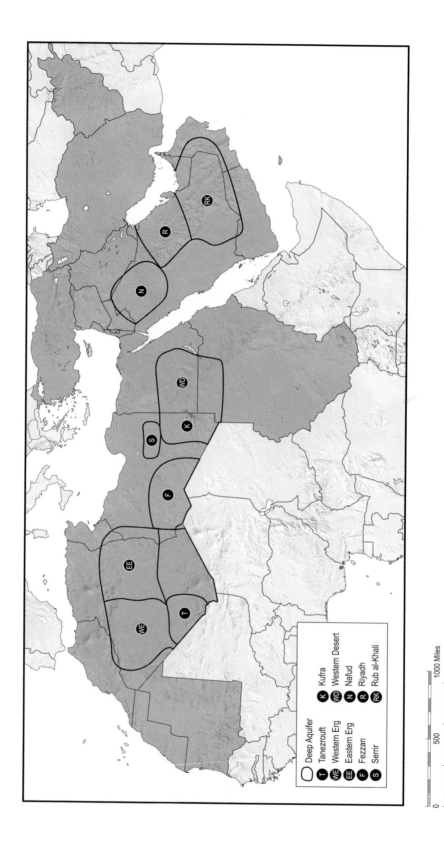

Map 8 Deep aquifers (boundaries approximate)

Deep Aquifer

T	Tanezrouft	K	Kufra
WE	Western Erg	WD	Western Desert
EE	Eastern Erg	N	Nafud
F	Fezzan	R	Riyadh
S	Serrir	RK	Rub al-Khali

1000 Miles

1500 Kilometres

500

500   1000

0   500

absent. By far the most extensive are lithosols, essentially weathered rock material with little or no humus.

Vegetation is controlled by many of the same factors that govern soil development except that soil acts as an interface between climate and the geological parent material. Organic factors include competition between plant types, grazing by animals and the often catastrophic influence of human activity particularly through vegetation clearance. As with soils, the vegetation pattern relates most obviously to the map of climate, modified by a number of variables. However, there are two key threshold values which mean that vegetation variation is more clear-cut. The upper threshold is 300–350 mm and the lower threshold is 80–100 mm of rainfall. Below the lower threshold rainfall-dependent vegetation can only exist if it is specially adapted to the conditions or if it is opportunistic and able to condense its life cycle into short wet periods. Between the two thresholds, the predominant vegetation will be scrub on the wetter margins, degenerating into increasingly sparse grassland. Above the upper threshold, open woodland and, in wetter areas, forest of various types predominates.

Human landscape

The physical landscape or system is overlain by and interrelates with the socio-economic system. In the Middle East, the predominant economic activity in terms of employment and geographical coverage is agriculture. However, for the region as a whole the economic focus is upon petroleum. While every state has agriculture, only a limited number have oil and natural gas in abundance. There is, therefore, a marked disparity in gross domestic product between the oil-rich and the oil-poor countries of the Middle East. Both agriculture and petroleum extraction have a marked effect upon the cultural landscape but it is only the latter which has generated sufficient controversy to be considered one of the key affairs of the Middle East.

Other than where irrigation intervenes, the map of agriculture reflects closely the map of rainfall. Broadly speaking, in areas with less than 100 mm, there will only be very scattered pastoralism, the density being always very low and varying from season to season and year to year. Between 100 mm and 400 mm, there will be what can be classified as rough grazing and this will be utilised by nomadic pastoralists. With rainfall of over 400 mm dry farming and the production of crops such as cereals becomes possible. With over 750 mm, depending upon altitude, Mediterranean polyculture can prevail. Irrigation, which facilitates intensive agriculture, is concentrated along the valleys of the Nile, the Tigris and the Euphrates and is relatively widespread in the western areas and Caspian coastlands of Iran, which has by far the highest area under irrigation of any state. Significant areas of irrigation, mostly scattered in pockets, occur from Morocco to Tunisia, particularly in the coastlands, along the southern part of the Levant coastline, in northern Yemen and in Turkey. Thus, throughout the core of the region, including central Iran, other than in oases, there is virtually no agriculture except for occasional extremely limited pastoralism and that is associated with oases.

The petroleum industry, producing oil together with gas and gas condensate, is located in two broad coastal belts (Map 9). One follows the line of the Gulf of Oman and the Gulf through the Tigris and Euphrates valley of Iran and Iraq. The other extends along the southern Mediterranean coastline from Egypt to Algeria. On the western side of the Gulf, from southern Oman through the UAE, principally Abu Dhabi and to a lesser extent Dubai, through Qatar, Saudi Arabia, Kuwait, Iran and Iraq, there is a belt predominantly of oilfields but with several key sources of gas, notably in Abu Dhabi, Saudi Arabia and Qatar. On the eastern side of the Gulf the belt extends from the Strait of Hormuz northwards through Iran and Iraq and is dominated by gas fields. Both belts have extensions most notably in Saudi Arabia and Iran.

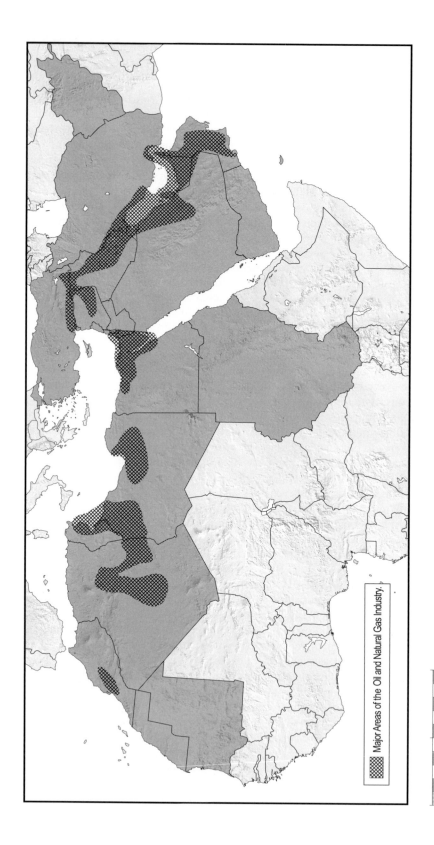

Map 9 Major areas of the oil and natural gas industry

Along the North African coast, oil predominates at the eastern end and gas at the western end, particularly in Algeria. The coastal distribution ends essentially at Tunisia as the majority of Algerian production is located inland. Outside these belts, there is production of oil in Turkey, and oil and gas in Syria, Israel, Morocco, Afghanistan and Yemen but these are all essentially on a limited scale. It can be seen that the core area of the Middle East, the great swathe of deserts is, unlike agriculture, significant in the production of petroleum products. The industry and its political implications are considered in greater detail in Section C.

Manufacturing industry illustrates a fundamental difference between the countries of the Middle East. A distinction can be made between those with long industrial histories and those reliant almost totally upon petroleum. Among the major oil-producing countries, there has been diversification both downstream in petroleum products and also into a variety of other industries. Turkey has the widest range of industry which varies from traditional crafts and textiles to modern engineering. Iran and Egypt in particular have followed suit. The state which has depended most heavily upon research and development and the use of technology is Israel. It has developed a number of high-grade industries aimed principally at niche markets. Over the last few years there has also been a significant increase in service industries particularly in Bahrain, the UAE and Lebanon. In the UAE, Dubai has grown as a regional trade and premier financial centre. In the context of the cultural landscape the distribution of industry: traditional, modern and service; relates closely to the distribution of population.

Leaving aside the location of major areas of irrigation and large parts of the petroleum industry, the maps of higher rainfall incidence, higher quality agriculture, manufacturing and settlement very largely coincide. The map of settlement is, of course, extended to include the irrigated areas and the petroleum industry. Therefore, it provides in a sense a summary of the location of economic activity (Map 9).

In the Middle East, there are two extensive areas of high population density with totals of over 100 inhabitants per km²: the coastline and immediate hinterland of the Levant from Aleppo to the Egyptian border and the valley of the Nile to well south of Khartoum (Map 10). Elsewhere, there are strips of similar density along the Tigris Valley, to the north of Baghdad, at the eastern end of the Turkish Black Sea coastline and along the Iranian coastline of the Caspian Sea. In addition, several of the major cities spread over a significant area, notably Casablanca/ Rabat, Algiers, Riyadh, Tehran, Tabriz and Ankara. If a statistic of over 40 people per km² is used, the areas listed can be extended to include European Turkey and the entire Turkish eastern and northern coastal belts, the northern border of Iran together with a corridor down to the head of the Gulf, most of the Euphrates Valley, virtually the entire coastline of Morocco, Algeria and Tunisia, sections of the coast of Libya around Tripoli and Benghazi, a large part of northern Yemen and a thin coastal spread, with notable gaps, from Muscat to Kuwait. As would be expected, there is little settlement in the arid core of the Middle East other than in connection with the petroleum industry and water sources such as rivers and oases.

The communications network reflects closely the map of settlement. The densest network of railways is throughout Turkey and the countries of the Levant with a rather more skeletal covering in Iran. There is also a pattern of railways along the coastlands of Morocco, Algeria and Tunisia. The other significant lines follow the valleys of the Nile, the Euphrates and the Tigris. The road network is dense throughout the most settled areas and sparse throughout the desert core. The main densities are in Turkey, the Levant countries, Iran and Iraq together with the northern parts of Morocco, Algeria and Tunisia. The other area with a significant density of roads is Sudan, south of Khartoum. The highest density of ports is related to the petroleum industry and is found round the Gulf. These include not only major ports but also single buoy moorings. Secondary densities occur along the Suez Canal and in the area of Alexandria, the Gulf of Sidra and Sfax in Tunisia. Densities of more traditional ports are located along the west

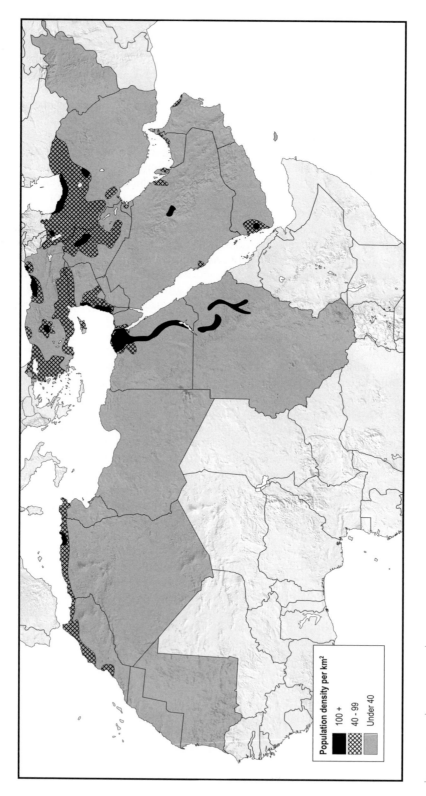

Population density per km²

■ 100 +

▨ 40 - 99

▒ Under 40

Map 10 Population density

coast of Turkey and the coastline from Morocco to Tunisia. Air routes show a strong focus upon the capital cities of the Middle Eastern states. In addition, Aden, Jeddah and, to a lesser extent, Casablanca, Dubai and Sharjah are significant hubs. While the pattern reflects that of population, given the problems of transport, it is unsurprising that the desert core of the Middle East is also sprinkled with minor airports.

Of particular significance for Middle Eastern affairs is the fact that the socio-economic system is overlain by the cultural system. There are many elements of this but the two which are most obvious and significant are language and religion. With both, but particularly in the case of the former, the difficulties of producing a map on the scale of the complete Middle East are acute. Many significant areas remain too small to be included while several of the major cities contain large minorities which also cannot be indicated.

Suffice it to say that the region is dominated by the Afro-Asiatic linguistic group which includes Arabs, Jews and Berbers (Map 11). This group is located throughout the African part of the Middle East other than part of southern Libya, sections of the Nile Valley and southern Sudan. It includes the entire Arabian Peninsula and the Tigris–Euphrates Valley to the fringes of Turkey and the Levant. The Turko-Altaic group, predominantly Turk, Turkoman and Azeri, includes virtually all Turkey, the Azeri portion of north-western Iran and the Turkoman border of north-eastern Iran. The remainder of Iran and most of Afghanistan is essentially Indo-European including Persian, Baluch, Kurd and Armenian. The sections of Middle Eastern Africa that are not basically Arab belong to the Nilo-Saharan group and include Teda, Nuer, Nubian and Dinka. In summary, the Middle East is in area overwhelmingly Arab but two of the three major states: Turkey and Iran, provide the core of separate ethno-linguistic groups.

With regard to religion, Islam is even more dominant than Arabic (Map 12). Only southern Sudan, with traditional Animist beliefs, Israel with Judaism and Cyprus with Christianity are entirely separate. There are also significant Christian minorities in Lebanon and, to a lesser extent, Egypt. The predominant branch of Islam is Sunni with more than four times the number of adherents claimed by Shi'a Islam. The Shi'ite focus is Iran, the only state which is officially of that persuasion. However, there is a Shi'ite majority in Iraq and Bahrain and there are sizeable minorities in Kuwait, Lebanon and Yemen. A further sect, Ibadism, is now only significant in Oman where it is the official branch of Islam. The importance of the cultural overlay is that so many issues in the Middle East are viewed through the prism of the particular language but more importantly Islam.

Finally, the human landscape is completed by the political overlay which is apparent in the mosaic of states and in a range of Middle Eastern affairs including the division of water sources and the settlement of boundary disputes.

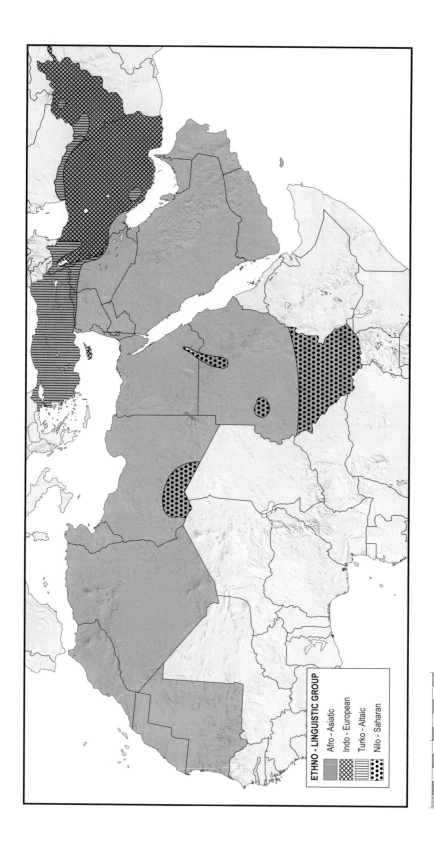

500
0
1000
500
0
1000
1500 Kilometres
500
1000 Miles

Map 11 Ethno-linguistic areas

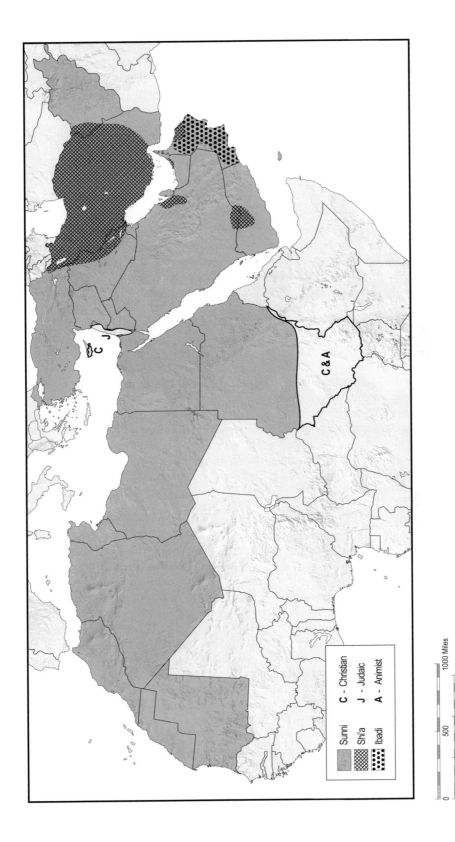

Sunni

Shi'a

Ibadi

C - Christian

J - Judaic

A - Animist

1000 Miles

500

1500 Kilometres

500   1000

0

0

Map 12 Religion

Historical background

To compress a history of the Middle East into so small a space is a daunting challenge. Arguably, civilised human history began in the area now known as the Middle East, and the region has been pivotal to the evolution of international relations ever since. More great empires have flourished and disintegrated in this region than in any other on the face of the globe. Although settlements in the area can be dated back as far as 10,000 BC, the earliest organised civilisations emerged from about 4,000 BC onwards. Centred in, or around the "land between two rivers" (the Tigris and the Euphrates) in what is now known as Iraq, a succession of highly sophisticated civilisations emerged to change the course of human history. The Sumerians were among the first civilisations to practise year-round intensive agriculture, based on highly developed systems of irrigation. By the late fourth millennium the Sumerians had also created a network of clearly demarcated city states and had developed what is considered by many to be the world's oldest written language. The Sumerians were superseded by a series of empires, the Akkadian, Babylonian and Assyrian, each of which left its own indelible imprint on the history of the region. The Akkadians, for example, developed a road network and postal service, while the Babylonians pioneered astrology and mathematics (including algebra and geometry) thousands of years before the ancient Greeks. Under King Hamurabi, the Babylonians also produced the world's first written code of law (Hamurabi's Codes) that established set punishments for specified crimes.

While Mesopotamia can justifiably claim to be the "cradle of civilisation", at roughly the same time (approximately 3,100 BC), the Nile River valley of ancient Egypt was unified under the pharaohs, ushering in an era of spectacular achievement in a variety of fields: construction, art, mathematics and medicine, until ancient Egypt was finally swallowed up by the Roman Empire in 30 BC.

The emergence of a significant power in the area now known as Iran dates from the sixth century BC with the rise of a series of Persian empires/states that culminated in the Achaemenid Empire led by Cyrus the Great. Under Cyrus and, subsequently, his son, Cambyses, the Persians conquered a large swathe of territory that encompassed much of the modern-day Middle East, extending as far west as Greece and Libya, and as far north as the Caucasus Mountains. The Achaemenid Empire was finally dismantled by a series of devastating military defeats at the hands of Alexander the Great and the encroaching Macedonian Empire. Persian empires would re-emerge, however, first in the form of the Parthian Empire in the second century BC, and subsequently, and more durably, in the form of the Sassanid Empire that lasted from AD 224 until AD 651.

Christianity took root in the Middle East as a consequence of the domination of the entire region by the Roman Empire. Christianity had spread widely throughout Roman lands and was officially adopted as the state religion of the empire by an edict of Emperor Theodosius I in AD 380. At the height of its powers, the Roman Empire united most of Europe and North Africa into a single political and economic unit. The empire was permanently divided into West and East in AD 395, and subsequently, the Western Empire fell in 476. The Eastern Empire, centred on Constantinople (formerly known as Byzantium) persevered as the Byzantine Empire and was a major power in the region until the fall of Constantinople in 1453 to Ottoman forces. For much of its lifespan, the Byzantine Empire was at war, either within itself, or, more destructively, with the rival Sassanid Empire. Beginning in 540, imperial rivalry between the Byzantines and Sassanids produced almost uninterrupted war between these two great powers until 629. With the economic weakness engendered by constant warfare against the Byzantines, the Sassanid Empire was ill-equipped to resist the onslaught of a powerful new military force that emerged out of the Arabian Peninsula from 632 onwards.

Rise and expansion of Islam

Muhammed ibn Abdullah was born in Mecca in 570 and led a mostly unremarkable life until close to his fortieth year. During what would subsequently be known as the "Night of Power", Muhammed received a direct message from God, via the angel Gabriel that he had been chosen as a vehicle of divine revelation. For the remainder of his life, first in Mecca, then Medina, Muhammed's received revelations were recorded into a single definitive text, the Quran, that was to form the basis of a powerful new religion that would subsequently transform much of the world. The first community formed on the basis of this new religion was at Medina in 622, and from these humble beginnings, the community had spread to encompass much of the Arabian Peninsula by the time of Muhammed's death in 632. Following Muhammed's death, prominent members of the community selected a successor (a "caliph") in order to protect the interests of the community. The first five caliphs, collectively referred to as the Rashidun (rightly guided) Caliphate, embarked on conquests that would shake the region to its foundations. Under the Rashidun Caliphate, Muslim armies emerged from the Arabian Desert to conquer Persia (633–51), Syria (637), Armenia (639), Egypt (639) and parts of North Africa (652). Under the subsequent Umayyad Caliphate, Muslim forces completed the conquest of North Africa, and by 750, the Arab Muslim empire stretched from the Iberian Peninsula in the west, to Afghanistan in the east, and from North Africa in the south, to the Caucasus Mountains in the north (Map 13).

In addition to this astonishing sequence of military conquests, the early Islamic period was also notable for the emergence of a major rupture in Muslim ranks. The split originated in a dispute over who would succeed Muhammed as leader of the community following his death in 632. The vast majority of Muhammed's followers favoured the caliph system whereby successors would be selected by prominent members of the community (the Sunnis), but a small percentage believed the succession should pass to Muhammed's closest blood relative, Ali (the Shi'a). This basic disagreement created the Sunni-Shi'a schism that continues to resonate to the present day.

End of Arab expansion

The sustained period of relentless Arab Muslim territorial expansion came to a sudden end during the eleventh century. From the north-east, Seljuk Turks migrated from their homeland in Central Asia to challenge Arab dominance throughout the Middle East. After conquering Persia, the Seljuks captured Baghdad in 1055, then moved on to Syria, Palestine and the Hijaz. Only Egypt resisted for any length of time, before falling to Turkish forces in 1169. Simultaneously, as the Byzantine Empire struggled to repel rampant Turkish forces, the first European Crusaders arrived

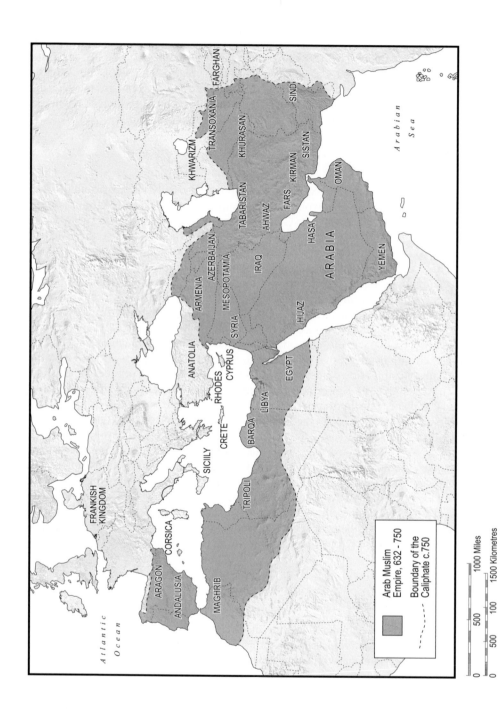

Map 13 Arab Muslim Empire (632–750)

in the Middle East to provide assistance. Landing in Asia Minor in 1079, the first Crusaders captured the coastal area of the Levant, from Cilicia to central Palestine. Responding to the pleas from Pope Urban II to recapture the Holy Land, the Crusaders dutifully captured Jerusalem in 1099. Jerusalem was in Christian hands until its recapture by the forces of Saladin in 1187.

Meanwhile, invading from the east, the Mongol hordes of Ghengis Khan presented a formidable challenge to established powers in the region. Between 1220 and 1227, Mongols attacked Iran repeatedly and destroyed most of its major cities. In 1258, Hulagu Khan, grandson of Ghengis, captured Baghdad and largely razed it to the ground, destroying in the process the Abbasid Caliphate and killing hundreds of thousands of the city's inhabitants. At the height of its power in the late thirteenth century, the Mongol Empire, stretching from the Sea of Japan to the Danube, held sway over more than 100 million people and covered over 20 per cent of the Earth's land surface (Map 14). It was supported by a network of land and sea trade routes (Map 15).

Rise of the Ottoman Empire

Named after its first great leader, Osman I, the Ottoman Empire began life as a small independent emirate in western Anatolia, but expanded its influence rapidly as a result of its unrivalled military prowess and the early incorporation of gunpowder into its military strategy. The founding of the empire is often dated from 1299, and during the century after Osman's death, the Ottomans won a series of decisive military victories in the eastern Mediterranean and the Balkans. In 1387, Ottoman forces captured Thessaloniki from the Venetians; in 1389, a comprehensive victory at the Battle of Kosovo (the so-called "Field of Blackbirds") extinguished Serbian power in the region; and in 1453, Turkish forces under the command of Mehmed the Conqueror captured Constantinople, which subsequently became the capital of the Ottoman Empire. During the centuries that followed, the empire solidified and expanded its control over southern Europe and the eastern Mediterranean. Under Sultan Selim I, the Persian Safavids were decisively defeated at the Battle of Chaldiran in 1514, and in 1517, Ottoman forces conquered Egypt. Sultan Suleiman the Magnificent, who governed the empire from 1520 to 1566, continued the military expansion of the empire, capturing Belgrade in 1521, and the Kingdom of Hungary in 1526. Expansion further into Europe was probably only prevented by the failure of the siege of Vienna in 1529 and the Great Siege of Malta in 1565. By 1566, the end of Suleiman's reign, the empire's population had reached somewhere in the region of 18 million people. However, the death of Suleiman also brought to an end the period of massive imperial expansion (Map 16). Over successive centuries, the Ottoman Empire was increasingly challenged by a resurgent Europe, fuelled by the wealth of the New World and growing technological prowess. By 1700, the Ottomans had been driven out of Hungary and during the Balkan Wars of 1912–13, the empire lost control of all its European possessions with the exception of Constantinople.

Elsewhere, the Ottomans suffered military setbacks throughout the Middle East. The French took control of Algeria (1830) and Tunisia (1878), while the British occupied Egypt (1882). Subsequently, as European powers expanded their reach into North Africa and the Gulf, the Ottoman Empire entered a period of terminal decline. The deathblow to the Ottoman Empire came as a result of its alliance with Germany and Austria-Hungary during the First World War.

Post-First World War

The First World War marked a decisive turning point in Middle Eastern history. During the war, in an effort to undermine the Ottoman Empire from within, the British encouraged the empire's

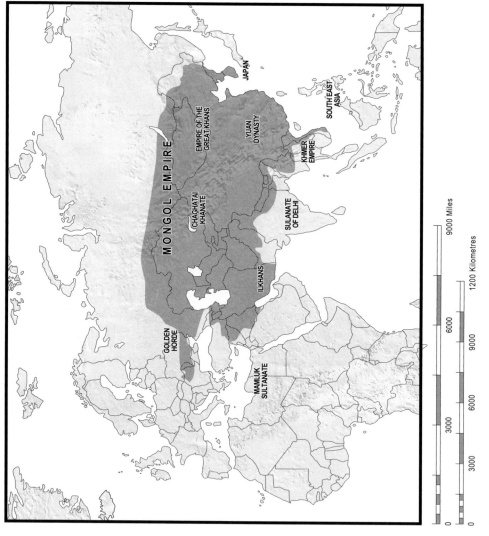

Map 14 Mongol Empire

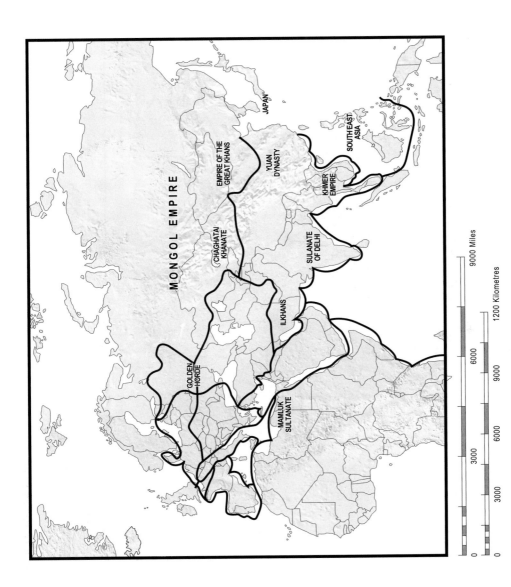

Map 15 Major trade routes in the Mongol Period

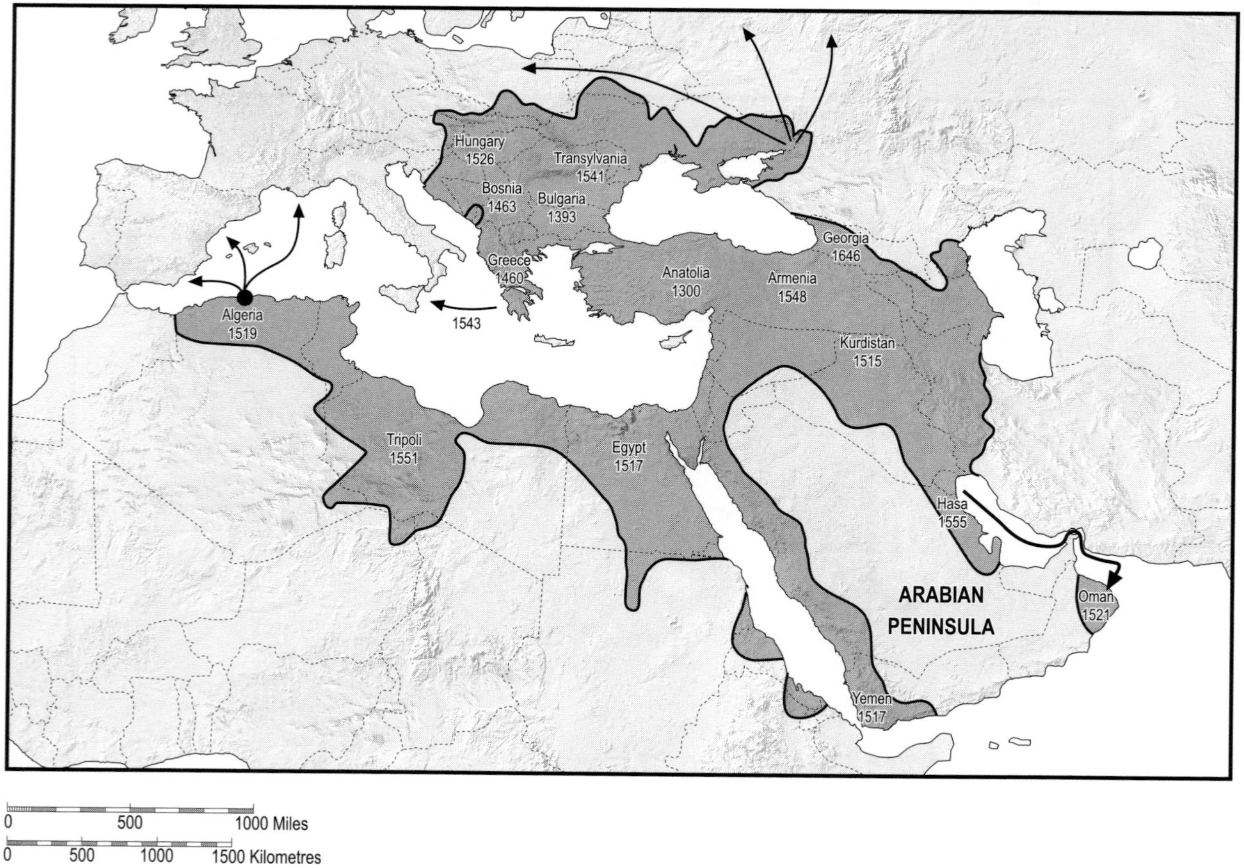

ARABIAN
PENINSULA

Oman
1527

Hasa
1555

Kurdistan
1515

Georgia
1646

Armenia
1548

Yemen
1517

Anatolia
1300

Egypt
1517

Transylvania
1541

Bulgaria
1393

Hungary
1526

Bosnia
1463

Greece
1460

1543

Tripoli
1551

Algeria
1519

1000 Miles

1500 Kilometres

0   500
0   500   1000

Map 16 Ottoman Empire (1663)

Arab subjects to rise up in revolt against Ottoman rule. The Arab Revolt, led by Sherif Hussein, helped hasten the demise of the Ottoman Empire and did much to fuel the emergence of Arab nationalism in the region. In return for their assistance in defeating the Turks, the British had promised to recognise an independent Arab state in the aftermath of the war. This promise was made explicit in a letter of 1915 from British High Commissioner in Egypt, Sir Henry McMahon, to Sherif Hussein, which pledged that, subject to certain modifications, "Great Britain is prepared to recognize and uphold the independence of the Arabs in all the regions lying within the frontiers proposed by the Sharif of Mecca."

Less than a year later, however, an agreement was signed by French and British diplomats, Francois Georges-Picot and Sir Mark Sykes, ever after known as the Sykes–Picot Agreement, that carved up large parts of the anticipated post-war remains of the Ottoman Empire into French and British zones of direct control and zones of influence. As shown on Map 17, Britain was granted control over areas corresponding to present-day Jordan and most of contemporary Iraq. The British also acquired control of the ports of Haifa and Acre. The French received control over a swathe of southern Anatolia, the Mosul region of northern Iraq, and areas roughly corresponding to present-day Lebanon and Syria. Imperial Russia, which assented to the agreement, was allocated control of Constantinople, the Turkish Straits and Armenia. Palestine was to come under international administration. Though signed in May 1916, the agreement was not made public until November 1917, the same month, indeed, as the "Balfour Declaration", in which the British government expressed its "sympathy with Jewish Zionist aspirations", and viewed "with favour the establishment in Palestine of a national home for the Jewish people". Sykes–Picot and the Balfour Declaration were perceived by Arab leaders as major betrayals of promises made to the Arabs during the First World War by the British government, though in fact nothing in the Sykes–Picot Agreement contradicted the idea of either single or multiple independent Arab states in the relevant region.

These behind-the-scenes deals made by the major powers about the design of the post-war order in the Middle East were superseded by the resurgent forces of Turkish nationalism under Kamal Ataturk, and the League of Nations and its associated mandate system. Under the system, in which Middle Eastern territories were to be "prepared" by great powers to assume the responsibilities of independence, the French gained the mandate for Syria and present-day Lebanon, and the British assumed responsibility for Iraq and Palestine. Faisal, one of Sherif Hussein's sons, was awarded the throne of Iraq, though he had never actually visited Iraq prior to his coronation, while the British Mandate of Palestine was divided into two territories. The eastern part emerged as the Kingdom of Transjordan, to which another of Sherif Hussein's offspring, Abdullah, was appointed king, and the western part was placed under direct British control as Palestine. The one part of the Middle East in which Arab aspirations for independent statehood came close to being achieved was in the Arabian Peninsula, where a British ally, Ibn Saud, eventually forged the Kingdom of Saudi Arabia in 1932.

The Interwar period

The future of the Middle East was irrevocably altered between the two world wars when the scale of the region's oil resources became evident. Oil had first been discovered in Persia as far back as 1908, and by the late 1920s, Iraqi oil was being pumped from the Kirkuk oilfield in the north, a megafield. Saudi oil was discovered in the late 1930s, and subsequently, smaller reserves were found in Algeria, Libya and Oman. With the world's largest and most easily accessible reserves of oil, the Middle East became indispensable to the Western world as a source of the material essential for industrialisation and economic development. Henceforth, control over the region acquired an unprecedented level of strategic significance.

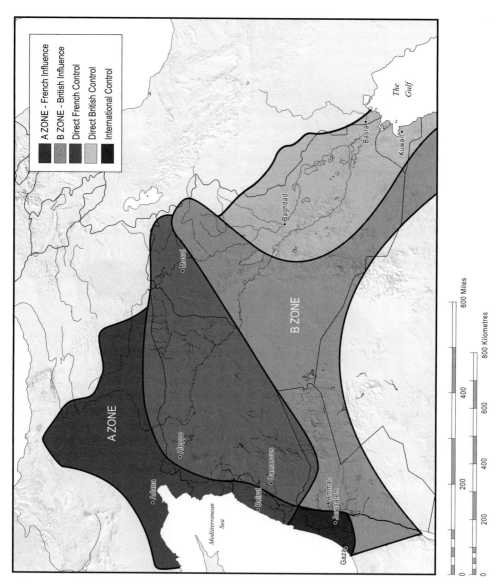

Map 17 Sykes–Picot Agreement (1916)

The other major development in the region during the interwar and immediate post-Second World War period was the formal emergence of most states from colonial domination. Iraq formally acquired independence in 1932 with the end of the British mandate, but the British maintained military bases beyond this point and exerted significant control over Iraq's foreign policy. Most Iraqis date real independence from the overthrow of the detested British-backed monarchy in the 1958 revolution. The other dates at which the states of the Middle East achieved independence are listed in the description of individual states in Section D.

Post-Second World War

The ravages of the Second World War left Europe's colonial powers gravely weakened and incapable of projecting power in the region. In the face of severe British weakness and with the expiry of Britain's mandate over Palestine looming, the United Nations (UN) attempted to forge a compromise partition plan to accommodate Palestine's Jewish and Arab population. The plan was rejected outright by the Arabs, and when the mandate expired in 1948, Zionist leaders unilaterally declared the state of Israel. This declaration triggered an invasion by neighbouring Arab states, and in the resulting Arab–Israeli War of 1948, Israeli forces succeeded in driving out large numbers of Arab inhabitants of Palestine, significantly expanding the borders of the nascent Israeli state. This was the first of three major wars between Israel and various Arab forces over the course of the next twenty-five years. Collectively, these wars resulted in the significant expansion of the borders of the Israeli state and created up to 1.5 million Arab refugees, many of whom live in squalid refugee camps to the present day. While Turkey recognized an independent Israeli state as early as 1949, the first Arab state to recognize the existence of Israel officially was Egypt as a consequence of the Camp David Accords and the Israel–Egypt Peace Treaty of 1979. Subsequently, Jordan signed a peace deal with Israel in 1994 and remains, along with Mauritania, the only other Arab country to conduct full diplomatic relations with Israel.

The retreat of European powers from direct involvement in the Middle East became inevitable following the debacle of the Suez Crisis in 1956. A joint British–French and Israeli assault to capture the Suez Canal, recently nationalised by the nationalist Egyptian leader Gamal Abdel Nasser, was initially very successful militarily, but floundered when the USA intervened to force an end to hostilities. Many people view this as the moment at which power in the Middle East was transferred from the European powers to the two new Cold War superpowers. During the Cold War, the strategic significance of the region gave rise to an intense struggle for influence and clients in the Middle East between the Soviet Union and the USA. Some states, such as Syria, were firmly in the Soviet camp until its demise in 1991, but the majority of states in the region were aligned with the USA for most of the Cold War period. Both Egypt and Iraq changed allegiances at various points, while Iran was an important US client until the Islamic Revolution of 1979 brought to power a regime that detested both sides.

During this period, US policy in the region was shaped by three key considerations: keeping the oil flowing westwards, protecting Israel against its Arab neighbours, and preventing the emergence of nationalist, anti-Western regimes that favoured nationalised control over oil resources and opposed the imperialist West and Israel as its Middle Eastern manifestation. The extent to which these three goals were mutually incompatible became evident in 1979 when the USA faced the prospect for the first time of being shut out by the Gulf's two main powers, Iran and Iraq. Iran came under the sway of a fundamentalist Shi'a regime bent on exporting violent anti-Western (and anti-Israeli) revolution throughout the Middle East. Iraq, under Saddam Hussein's leadership, had signed a treaty of friendship with the Soviet Union as a prelude to nationalising the Iraqi oil industry in the early 1970s. The drawn-out Iran–Iraq War was, therefore, something of a boon to the USA, enabling the USA to support both sides at various stages of the conflict and to ensure that both were exhausted by the end of the war.

Post-Cold War

The demise of the Soviet Union in December 1991 left the USA as the region's, and indeed, the world's, one remaining superpower. An early indication of the power of the USA was its capacity to mobilise a large coalition of forces, including several Arab countries, under the UN banner to drive Saddam Hussein's troops out of Kuwait. The 1991 Gulf War was an awe-inspiring display of US soft and hard power that resulted in a shattering defeat for an army that had been assumed to be among the most powerful in the region. Subsequent to the war, the USA spearheaded the campaign to disarm Iraq of its weapons of mass destruction (WMD) and kept stringent economic sanctions in place until 2003. The sanctions had a devastating effect on the social and economic life of Iraq, reducing the living standards of the population to sub-Saharan levels, but did nothing to weaken Saddam's hold on power. The Clinton administration tried to capitalise on the US position as the sole remaining superpower to reach a peace deal between the Israelis and Palestinians. Though some progress was made, a final deal over contentious issues such as the status of Jerusalem and the right to return for refugees proved elusive.

September 11 and its aftermath

The attacks on the Pentagon and World Trade Center of 11 September 2001 (9/11) demonstrated to many that US policy towards the Middle East was in need of a radical overhaul. The administration's initial response, an attack on the Taliban regime in Afghanistan, was generally viewed as a proportionate and justified response to the attacks of 9/11. It soon became clear, however, that the response to 9/11 would be much more ambitious. In early 2002, President George W. Bush identified an emerging "axis of evil" comprising North Korea, Iran and Iraq, which collectively constituted a gathering threat to the USA. The qualifications for membership of the axis were never made entirely clear. The administration's subsequent National Security Strategy (NSS) justified a doctrine of pre-emptive (though actually preventive) war that would allow the USA to strike at enemies before the nature of the threat had fully materialised. The NSS also emphasised the importance of the spreading of democracy as a means of fostering peace in the Middle East. Iraq was to be the first concrete application of this radical new doctrine. Though the administration struggled gamely to provide tangible evidence of operational links between Saddam Hussein's regime and Al Qaeda, and strongly implied the existence of compelling evidence on the existence of ongoing Iraqi WMD programmes, the USA was unable to convince many that Iraq constituted a clear and present danger to international security. President Bush's "coalition of the willing" was a shadow of the imposing military force painstakingly pieced together by his father in 1991.

A US/British force of approximately 200,000 finally invaded Iraq in March 2003, and the ensuing Operation Iraqi Freedom lasted approximately three weeks before Baghdad fell to coalition forces. The failure of the administration to unearth the promised WMD led to a change in emphasis regarding the motives for the invasion. Increasingly, the Bush administration's post-hoc rationale was the desire to spread democracy throughout the Middle East as a means of addressing terrorism's root causes. Iraq was to provide the trigger for a democratic tsunami that would clear the region of tyrannical rulers, many of whom had been actively supported by the USA at one stage or another during the Cold War, and liberate the peoples of the Middle East. The comprehensive mismanagement of the post-war situation in Iraq raised important questions about the capacity of the USA, with so few troops on the ground, to deliver security, let alone democracy, to the Iraqi people. The occupation of Iraq soon degenerated into a bloody quagmire for US forces, periodically punctuated by brief moments of hope, such as the elections of January 2005 and December 2005. The so-called "surge" of 2008 was thought at least partly responsible for reducing violence levels significantly over previous years', but a lasting solution to Iraq's underlying political problems probably became more elusive as a result. In particular, funding Sunni Arab militias to drive out Al Qaeda may be effective in the short term, but it is difficult to see how empowering disgruntled Sunni militias that despise the Shi'a government in Baghdad will contribute to Iraq's stability in the long term.

Elsewhere, the Bush administration's goal of promoting democracy throughout the Middle East confronted a brick wall in states such as Egypt and Saudi Arabia. In the Palestinian territories and Lebanon, the strategy backfired in spectacular fashion. A stunning victory for Hamas in the election of 2006 and the growing political power of Hezbollah in Lebanon provide ample indication that, given US current core interests in the Middle East, promoting democracy is likely to undermine rather than enhance the US strategic posture in the region.

As of 2008–9, the Middle East is in a state of flux. Newly unleashed forces, such as the political power of organised Shi'ism, and the military power of sub-state actors, such as Al Qaeda and Hezbollah, cohabitate uncomfortably with the long-standing geopolitical concerns, such as the Israeli–Palestinian conflict and the simple fact that the Middle East contains the world's largest reserves of fossil fuels. Predicting the future is, therefore, as challenging as summarising the past.

Section C
Fundamental concerns

As the settlement pattern results from a combination of geography and history, mostly on a relatively local scale, so the mosaic of states offers a similar summary on a regional scale. Each state comprises territory and an organised population. To exercise control over territory, the limits of that territory must be known, and therefore international boundaries are of great significance. This fact can be confirmed worldwide when it is realised that some 70 per cent of international conflict relates in some way to a boundary concern. Despite increasing globalisation, the states remain important as they alone can help enforce world order. The states also control the resources within their territories. There is a range of resources throughout the Middle East, but the most crucial by far are water and petroleum.

Very few Middle Eastern states have no water problems and those which are more or less problem free, such as Lebanon, are still implicated in regional hydropolitics. The petroleum industry pervades the entire region and no country is immune from the economic and political vicissitudes that result. If maritime boundaries are included, it is clear that no Middle Eastern state is without some sort of boundary issue. All are concerned to some degree with transboundary problems, predominantly the trafficking of drugs and people.

These fundamentals: petroleum, water, international boundaries and transboundary movement; influence to a greater or lesser extent the behaviour of all the Middle Eastern states as actors upon the world stage. It is appropriate, therefore, that they should be considered before the states and separately from Section E in which key issues concerning the behaviour of certain specific states are discussed.

Petroleum

Resources entered the realm of politics with the dawn of the oil age. Previously, before the early part of the twentieth century, coal had been the dominant fuel mineral. Industrialised countries relied upon their own or nearby resources and very little coal entered world trade. Global trade relied upon shipping and ships were fuelled by coal. To ensure supplies, bunkering points were established along the main sea-lanes at ports such as Aden, but fuel mineral vulnerability was never considered a serious threat. Furthermore, coal is distributed far more evenly globally than oil and, in emergency, alternative supplies would have been readily available. Also, it must be remembered that the Great Powers controlled the world order far more tightly at that time.

The situation changed dramatically as oil became the major energy source, first for shipping and then for industry. The developed world became increasingly dependent upon a limited number of developing countries for oil. Many of these countries are unstable, although it has to be recognised that such instability is commonly related to the possession of oil. As the number of key suppliers has always been limited, the links between them and the consumers have become strategic routeways, adding to the overall vulnerability. Particularly at risk are pipelines since their location is fixed. Pipeline diplomacy was illustrated during the Gulf War (1990–1) when Turkey switched off Iraq's main export pipeline to Dortyol.

Therefore, pipelines offer scope for both legal and illegal interference. In fact, terrorist activity in blowing up pipelines is likely to have a limited effect since the supply of oil or gas can be switched off at many points along the pipeline. Furthermore, pipelines may be protected in various ways and do not offer as simple a target as is commonly supposed. However, pipelines both internal and transboundary do present a risk. During the lengthy unconventional phase of the Iraq War, following the conclusion of formal military action, the destruction of pipelines has been used as a method of putting pressure upon the government and other authorities. With transboundary pipelines, the risks are greater since control and security are shared and it is more difficult to attribute blame for deliberate damage. Terrorism which destroys pipelines can result in international problems.

The Middle East has a dense network of oil, gas and products pipelines which links the petroleum sources to the processing centres, ports and cities. There is also a growing framework of transboundary pipelines, particularly apparent in the links between the *Maghrib* and Western Europe and from the Caspian basin into the states of the Middle East (Map 18).

Shipping routes must also be considered as a potential source of disruption to supplies. The exact position of sea-lanes is dictated by both economic and physical conditions with the result that there is a tendency for them to converge around promontories and particularly through restricted channels. Areas where sea-lane restrictions result in a concentration of shipping within a limited area are known as choke points. It must also be realised that navigation within choke points is usually further curtailed by bathymetry, islands, sandbanks and wrecks together with the normal activities of ferries, fishing boats and pleasure craft. Map 19 illustrates the concentra-

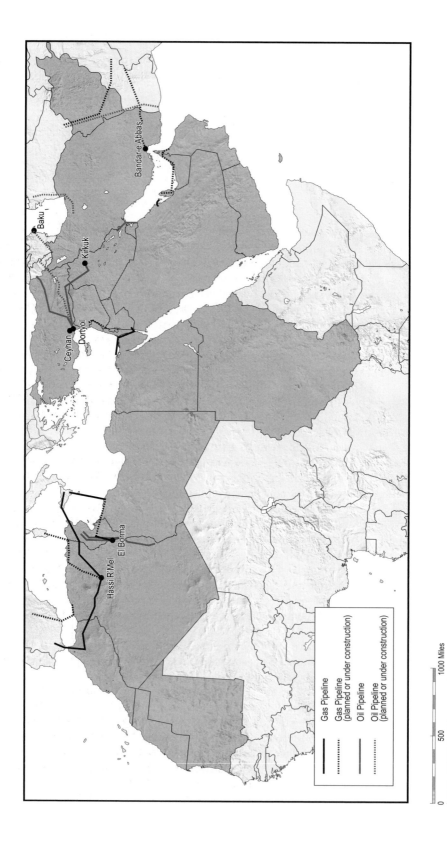

Map 18 Transboundary pipelines

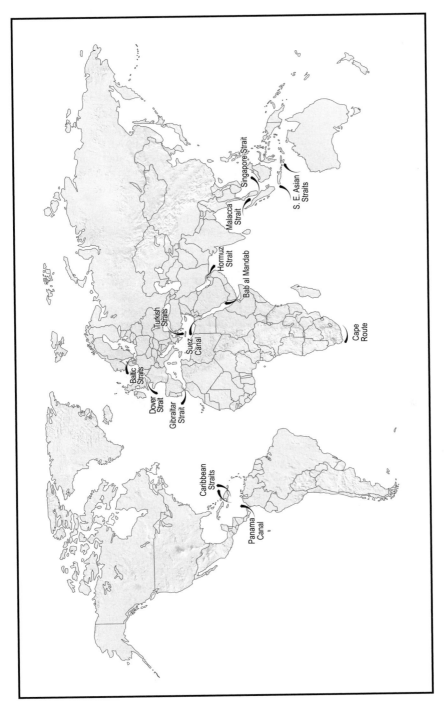

Map 19 Key choke points

tion of global choke points in the Middle East region. Given that they have the highest densities of merchant shipping in the world, choke points are an obvious location at which overt or covert interference can take place. Apart from the Malacca and Singapore straits, piracy is currently concentrated off the Strait of Bab al Mandab and the coastline of Somalia.

The most extreme case of sea-lane closure at a choke point occurred in the context of the Suez Canal. In 1956, the UK, France and Israel went to war against Egypt as a result of the nationalisation of the Suez Canal Company and the outcome was that the canal was blocked by shipwrecks until April 1957. In June 1967, following the invasion of Sinai by Israel, the canal was again blocked and was not reopened until June 1975. The result of the second closure was particularly far-reaching. Tankers had to transit the sea-lane round the Cape of Good Hope and the additional costs involved saw the advent of super-tankers; by 1975, on its reopening, only 27 per cent of the world's tanker fleet was small enough to navigate the Suez Canal. Finally, it must be remembered that choke points have been discussed entirely in terms of petroleum but they constitute a subject in their own right since they are of significance to all merchant shipping whatever the cargo and indeed to warships.

Oil is the only truly global industry, with a network of interrelationships which completely enmeshes the world. Since the key areas of production and consumption tend to be separate, oil is also the major commodity in international trade. There are now many focal points of this industry including the headquarters of the major oil multinationals and the stock markets of New York, London and Tokyo but the major focus, of increasing importance, is the Gulf. The development of the oil industry in the region has always been a power struggle and therefore has always involved politics. The ascendancy passed from the major oil companies, initially the "Seven Sisters", later joined by the "Independents", to the producer countries themselves, then to the Organisation of Petroleum Exporting Countries (OPEC) and, finally, to the entrepreneurs and the stock market. The effects of the long period of colonialism and Byzantine intrigue are still felt in the region today.

The word "concession" is still highly emotive since it involved the establishment of some very one-sided partnerships. Indeed, much of the residual anti-Western feeling in the region stems from the period when the producer countries fought to free themselves from the iron embrace of the international oil corporations. The concessions, as in the cases of Kuwait, Qatar and Bahrain, often covered the entire country and the specified duration was usually up to seventy-five years. The companies had virtually complete freedom and enjoyed a wide range of privileges which rendered them free from local government control. As a result, the benefits to the Middle Eastern producer countries were very limited.

The injustices were gradually curtailed and the balance finally changed after the nationalisation of the Anglo-Iranian Oil Company's concession in 1951. Accommodation to a 50–50 share-out between the new consortium, comprising British and American interests, and the Iranian government was reached in 1954. The new system was best exploited by Libya, a new arrival on the scene, which awarded concessions, comprising fifty-one separate areas, to seventeen oil-producing companies, many of which were comparatively small. The concessions were granted for relatively short periods and the role of the corporations was thereby forcibly changed from that of a concessionaire to that of a contractor. The move towards a fairer return on resources for the producer countries was given further impetus following the creation of OPEC in 1960. The establishment of OPEC signalled the fact that economic independence was to be congruent with political independence. This was demonstrated during and after the Arab–Israeli War of 1973 when two substantial price increases were imposed. However, the price then stood at $11.65 per barrel, hardly excessive in the light of the enormous profits garnered by the corporations.

The result was a massive transfer of wealth to the producer countries which, with their low absorptive capacities, invested heavily in Western banks. At that time, therefore, the Middle

Eastern producers controlled not only the oil but were also extremely influential in the world banking system and the full array of fiscal affairs of many Western countries. Since that time, there have been periods of relative price stability but also major high points. Oil peaked at $43 a barrel in 1980, at $40 in autumn 1990 and at $147 in July 2008. However, during that period power was further dispersed by the entry on to the scene of the Rotterdam market. This spot market allowed shortfalls and surpluses to be accommodated and the definition of the price of oil became the price at which these marginal barrels changed hands. Both OPEC and the oil corporations started pegging their contracts to the spot market.

During the 1980s, the corporations began to rely even more on the spot market and they eventually surrendered control of oil prices at the margin. The Wall Street "refiners" reigned supreme with the introduction of options and futures markets which brought new volatility into oil prices. The price of oil was thereby redefined as that quoted in the futures market rather than that traded at the present time. The trading system shared out risk and reward in a manner few lay-people understood. As a result, with the onset of the Gulf War in August 1990, oil prices increased rapidly, not as a result of shortage but of psychology.

Thus, the wheel has effectively turned full circle in that power is once again being exercised by those operating far from the oilfields. They may lack the manipulative opportunities of the major oil companies but their interests do not coincide with those of the Middle Eastern countries. It is a sorry saga of greed, collusion and perfidy which, interwoven with Arab nationalism and burgeoning Islamic Fundamentalism, has produced a series of Middle Eastern crises.

The Middle East has twelve mega-fields, those with reserves of over 1,000 million barrels of oil. These include two of the largest in the world: Ghawar in Saudi Arabia and Burgan in Kuwait. Iran has five such fields, Iraq two, Kuwait one, Saudi Arabia three and Libya one. In 2007 the Middle East was responsible for 36.8 per cent of world oil production. However, in considering the strategic nature of oil, the important figure is not the production but the potential to produce as shown by proved reserves. According to the British Petroleum *BP Statistical Review of World Energy* (June 2008), proved reserves of oil are generally taken to be those quantities that "geological and engineering information indicates with reasonable certainty can be recovered in the future from known reservoirs under existing economic and operating conditions". This definition indicates that the figure for reserves can change not only as a result of geological exploration but also as a result of advances in mining technology and global economic conditions. For example, in 1987 there was a major reappraisal involving Iran, Iraq, the UAE and Venezuela and global proved reserves increased by some 27 per cent. In the past twenty years, total world reserves have increased by 36 per cent. There is always likely to be some relationship between oil prices and proved reserves in that enhanced prices can produce further exploration and improved extraction techniques.

Proved reserves as for the end of 2007 are set out in Table 1. In 2007, the Middle East, the countries of the Gulf together with those of the North African littoral, totalled 66 per cent of global reserves while the Muslim world in general is estimated to have well over 70 per cent of reserves. The table lists all the countries with major reserves (2 per cent or over of world reserves) but also includes China and India as major consumers. Algeria, the second most important source in North Africa, has only 1.6 per cent of world reserves.

The pre-eminence of the Gulf States is very clear. It is enhanced further by a consideration of the reserves to production (R/P) ratio. This is defined in the *BP Statistical Review of World Energy* (June 2008): "if the reserves remaining at any year are divided by the production of that year, the result is the length of time that those remaining reserves would last if production were to continue at that rate". The R/P ratio gives an interesting view on the likely longevity of reserves but, since it depends on production rates which may vary with economic, technological and geological factors, it is a less accurate guide than the percentage. While in general the R/P

Table 1 Oil: proved reserves, 2007 and 2000

	2007		2000	
	% of world reserves	*R/P ratio*	*% of world reserves*	*R/P ratio*
Saudi Arabia	21.3	69.5	25.0	81.1
Iran	11.2	86.2	8.6	65.7
Iraq	9.3	Over 100	10.8	Over 100
Kuwait	8.2	Over 100	9.2	Over 100
United Arab Emirates	7.9	91.9	9.3	Over 100
Venezuela	7.0	91.3	7.3	66.4
Russian Federation	6.4	21.8	4.6	20.6
Kazakhstan	3.2	73.2	0.8	31.1
Libya	3.3	61.5	2.8	55.3
Nigeria	2.9	42.1	2.2	29.4
USA	2.4	11.7	2.8	10.4
China	1.3	11.3	2.3	20.2
India	0.4	18.7	0.4	17.3

ratios have declined slightly since 2000, the variable nature of the ratio is illustrated by Iran where it has increased from 65.7 to 86.2 and even the USA where there was a marginal increase from 10.4 to 11.7. It can be seen that the R/P ratio for China has declined steeply over the same period. The figures indicate that the Middle East will remain pre-eminent for the foreseeable future as the global source of oil while countries such as the USA and particularly China will require increasing imports. This situation illustrates clearly the potential for political conflict in the Middle East.

Figures for production (Table 2), taken in combination with those in Table 1 illustrate further the pre-eminence of the Middle East. At present, the Russian Federation is the second highest producer and the USA the third after Saudi Arabia. However, the R/P ratios of both indicate that this is likely to be a temporary situation. The Middle East as a whole, including smaller producers such as Sudan and Tunisia, is responsible for 36.8 per cent of global production, more than three times that of the Russian Federation. However, the increasingly dominant role in global energy production of the Russian Federation underlines further the growing political role of oil.

Overall there has been a general decline in consumption within the developed world although global consumption has increased every year over the past ten years. Indeed, since 1987 the total has increased by 15 per cent. The USA remains by far the world's largest consumer, followed by China which has increased its consumption by almost 88 per cent over the last ten years (Table 3).

The relationship between production and consumption shows any deficit which needs to be covered by imports (Table 4). This illustrates the growing deficit of China, but in particular, the parlous position of the USA. Thus, there is potential for competition between West and East over oil from the Middle East.

International trade in oil accounts for 64 per cent of consumption and since 2000 this trade has grown by 26 per cent. Almost two-thirds of the oil consumed worldwide has to be imported and there are at present five main exporting regions: the Middle East (42 per cent), the former Soviet Union (15.2 per cent), West Africa (8.8 per cent) and South and Central America (6.5 per cent). Again, the Middle East predominates as the key source for imports and, given the R/P

Table 2 Oil: production, 2007

	% of world production		% of world production
Saudi Arabia	12.6	Kazakhstan	1.8
Iran	5.4	Libya	2.2
Iraq	2.7	Nigeria	2.9
Kuwait	3.3	Algeria	2.2
United Arab Emirates	3.5	USA	8.0
Venezuela	3.4	China	4.8
Russian Federation	12.6	India	1.0
		UK	2.0

Table 3 Oil: consumption, 2007

	% of world consumption		% of world consumption
Saudi Arabia	2.5	Italy	2.1
Iran	1.9	Spain	2.0
Russian Federation	3.2	USA	23.9
Germany	2.8	China	9.3
Canada	2.6	India	3.3
France	2.2	UK	2.0
Mexico	2.3	Japan	5.8
		South Korea	2.7

Table 4 Import dependence, 2007 (million tonnes)

	Production	Consumption	Deficit
USA	311.5	943.1	631.6
Russian Federation	491.3	125.9	–
UK	76.8	78.2	–
China	186.7	368.0	181.3
India	37.3	128.5	91.2

ratios, this situation is likely to be enhanced in the future. Map 20 illustrates the current dependence on imports from the Middle East. The key fact is that 55 per cent of oil from the Middle East goes eastwards and only 34 per cent to Europe and the USA. Indeed, this latter figure is enhanced by supplies from North Africa and only 26 per cent of exports from the Gulf region go to Europe and the USA. Since 2000, the Gulf countries have increasingly favoured the main importers from the Asia-Pacific: Japan, China and Singapore. Given the long-term dependence upon the Middle East region, it is somewhat alarming to see the reorientation of its market in favour of the Pacific region rather than the West. Since 1989 and the eve of the Gulf War, US dependence on Gulf supplies has declined from 24 per cent to 16 per cent and European dependence has halved from 42 per cent to 21 per cent. Western reliance has to be placed increasingly on countries which have far smaller reserves and are politically less stable.

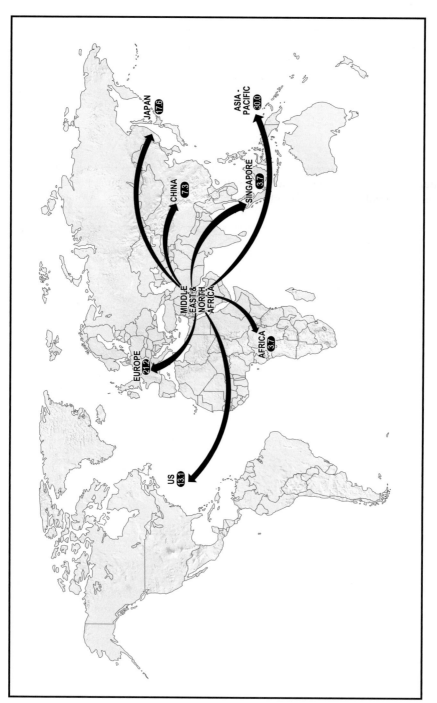

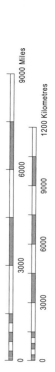

Map 20 Oil: percentage of total exports from the Middle East and North Africa (2007)

These figures illustrate the point that the major long-term source of global oil will be the Middle East and the current trading pattern is increasingly favouring the Pacific region. As the economies of China and India gather pace, this trend is likely to be accelerated. Therefore, resource politics is likely to play an increasing role in relations between Europe and America on the one side and India and the Asia-Pacific, particularly China, on the other.

There are a number of possible countermeasures. In the short-term, interruptions can be overcome by stockpiling. This would be accomplished through using oil already in the system but for severe disruptions it would need to be addressed through a national emergency petroleum stockpile. The US Strategic Petroleum Reserve provides fifty-nine days of import protection, 118 days if private stocks are included.

If shortages are likely to be longer term, attempts must be made to diversify suppliers but suppliers are limited globally and direct substitution from another source is not always possible without changes in oil treatment procedures. Conservation has proved effective to a certain degree but for transport neither conservation nor substitution provides anything like a complete solution. The main potential substitutes for oil remain other sources of energy. A major alternative, often found in association with oil, is natural gas. The problem is that natural gas is considerably less equitably distributed globally than oil. Over 55 per cent of reserves are shared by only three countries and those are also potential suppliers of oil.

Proved reserves of natural gas are set out in Table 5 and these show the complete dominance of the Russian Federation, Iran and Qatar. The remaining Gulf oil-producing countries also have significant supplies along with Algeria, Egypt and Libya. Therefore, in many ways the most obvious substitute for oil is likely to encounter the same problems as are found with oil itself.

Table 5 Natural gas: proved reserves, 2007

	% of world reserves	R/P ratio
Russian Federation	26.6	80.0
Iran	14.9	Over 100
Qatar	14.3	Over 100
Saudi Arabia	3.8	99.3
United Arab Emirates	3.4	Over 100
Iraq	1.8	Over 100
Nigeria	2.9	Over 100
Algeria	2.5	52.2
USA	3.0	10.4
China	1.3	47.0
India	0.6	36.2

Water

Aridity is basically defined as a lack of moisture and is therefore based upon a number of climatic variables. The exact boundary of the arid zone has long been a source of contention but in 1977 the United Nations Educational, Scientific and Cultural Organisation (UNESCO) published a map which is now accepted as standard (Map 21). The boundaries illustrated enclose both the arid and the hyperarid regions and include all the areas within which the mean annual precipitation divided by the mean annual potential evapotranspiration (both in mm) is less than 0.2. The main areas of hyperaridity are in North Africa and the Arabian Peninsula and the map confirms the fact that the Middle East is the major area of global aridity. As can be seen, from the map, it is not the only area of aridity but it is the most extreme for any area with such a relatively high population and such a crucial strategic resource as oil. The climate of the entire region, with the exception of the mountainous areas of the *Maghrib*, Turkey, Iran and Afghanistan together with southern Sudan, falls within the UNESCO classification of arid.

Water within the Middle East has been regarded as critical since the earliest times as is shown by the large number of major water engineering constructions, the remains of which can still be seen. The early civilisations in Mesopotamia and the valley of the Nile are thought to have developed because of the possibilities offered by the control of water. The necessity to share and distribute equitably the available water is thought to have initiated the development of many skills and sciences. The remains of ancient irrigation systems have been recorded at sites as old as 5,500 BC in Iran and 2,300 BC in Mesopotamia. Irrigation was introduced to the Nile in about 3,300 BC and Egypt claims to have the world's oldest irrigation dam which was constructed at about that time. The greatest of the ancient aqueducts, the Aqueduct of Carthage, was built in the second century AD, stretched for some 141 km and had a capacity of 31.8 million litres per day. The oldest canals, located near Mandali in Iraq, have been dated at about 4,000 BC. The longest canal in the Islamic world was the Nahrawan Canal in Iraq which had a total length of 380 km. Underground irrigation channels or qanats have been in use in the Middle East for over 2,000 years.

This preoccupation with water and the security of access to it continues in the region to the present day. Indeed, given the far greater requirements throughout the region, water is today an even more sensitive issue.

Precipitation is scarce and irregular and there is throughout vast tracts of the region only ephemeral surface flow. Between 18° N and 30° N, apart from the Nile, the two main sources of which rise in entirely different climatic zones, there is no permanent surface flow (Map 7). Therefore, the surface storage of water is limited to the major river basins and the wetter areas around the periphery of the Middle East. Whereas in humid climates the river and the lake are central to water management, in arid areas the underground water store or aquifer is crucial. The water system is based upon subsurface storage, much of it in relatively shallow aquifers. These are recharged by rainfall and snowmelt together with various artificial recharge procedures. In

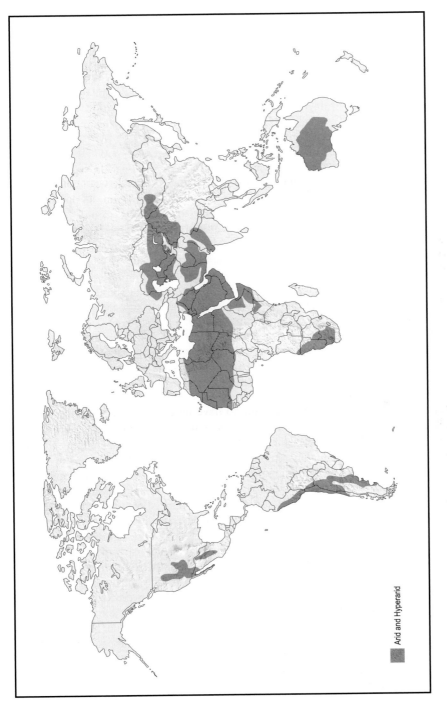

Arid and Hyperarid

Map 21 Arid regions of the world

the case of the deep aquifers (Map 8), it is not known whether any contemporary recharge occurs. When viewed as a system, it can be seen that there are irregular inputs to the aquifers but, in general, regular outputs. While precipitation is scarce and highly variable, evaporation, evapotranspiration, aquifer leakage and particularly human extraction cause regular depletion.

In such environments, water management is clearly crucial and poses complex problems. Water is a finite, mobile and vulnerable resource and, in the Middle East, conditions under which supplies need to be provided and distributed are as hostile as any on Earth. Water is the most versatile commodity and management needs to take into account its many functions, each of which has to be prioritised according to its relative importance, judged socially, economically and, most important, politically. If supplies are reduced, adjustments need to be made to address as many functions as possible without inciting conflict.

In the Middle East, the problem is exacerbated by the rapidly increasing population and urban development. Between 1900 and 2000, the population of the Middle East increased from 41 million to 400 million and the urban proportion of the population is now approximately 60 per cent, having been only 30 per cent in 1950. The key point about urbanisation is that nomadic rural people use between ten and twenty litres per person per day but in cities individual consumption may rise to 800 litres. Thus, in many of the Middle Eastern states, particularly those in the Arabian Peninsula, consumption far exceeds natural replenishment. Furthermore, given the problems of waste disposal in developing societies, aquifers are threatened with pollution. Longer term, there is the issue of rising sea level and the possible infiltration of saline water into the aquifer. Also, the water systems of the urban Middle East, in common with those of most of the Western world, suffer from leakages which are estimated to range between 30 per cent and 50 per cent of total supply.

In an attempt to address some of these problems, a variety of countermeasures involving either increases on the supply side or reductions on the demand side have been implemented. Approaches to increasing supply can be summarised in the following categories: (i) storage; (ii) research and development; (iii) the location of alternative sources. In the Middle East, storage is normally in the form of groundwater in aquifers but there is, of course, surface storage in the three international basins: the Nile, the Tigris–Euphrates and the Jordan. However, in each case, evaporation losses are extreme. It is possible to enhance groundwater storage by artificial recharge, for example by building recharge dams. There are hundreds of these, ranging from the very sophisticated to the simple manual emplacement of gabions, throughout the region. However, there are several problems including the fact that the dams are likely to be in use for very limited periods in the year and evaporation losses will be high. A major issue is sedimentation which greatly inhibits infiltration rates. Water can also be injected or spread into aquifers in an attempt to avoid the extremes of sedimentation. However, the fundamental issue results from the slow transmission rates which means that there is very little evidence linking recharge with the enhancement of aquifers which may be located some distance downstream.

The major source of additional water and the major result of research and development is desalination, the production of fresh water from brackish water or seawater. In the early 1960s, the multi-stage flash technique became available but this process has been overtaken by a variety of processes using membranes, particularly reverse osmosis. The Middle East dominates desalination with well over 60 per cent of the installed capacity in the Arabian Peninsula. Multi-stage flash still predominates but, apart from reverse osmosis, electro-dialysis and vapour compression desalination are being introduced either separately or in concert. Depending upon local conditions and the process used, operating costs can range from $0.5 to $1.3 per m^3. Therefore the costs of meeting, for example, the needs of Palestinians would be in the order of $100 million.

A second method of supply enhancement is through the recycling of waste water, chiefly from sewage disposal. Relatively few cities in the Middle East have adequate sewerage systems but recycling is receiving increasing attention and a number of countries including Israel, Jordan, Qatar, Kuwait and Saudi Arabia have significant recycling capacity. However, the effects on national water budgets are as yet limited.

In the Middle East, as is apparent from the map of precipitation (Map 6), alternative sources of water from neighbouring countries are limited. The one country with a reasonable excess is Turkey and in 1986 a Peace Pipeline was proposed to carry water from the basins of the Ceyhan and Seyhan on the Mediterranean coast to the states of the Red Sea and the Gulf. Various routes were planned and costed but ultimately, security issues prevailed and the pipeline was never built. Water from the pipeline was intended primarily only as a standby and the fact that agreement could not be reached between the various states indicates the strong psychological feelings about water security.

In 2002 the government of Israel made two fundamental decisions about water. The first was to double its production of desalinated water to 400 mcm/y. The second was to import 50 mcm/y from Turkey over a twenty-year period, the water being transported in converted oil tankers. Thus concluded a series of experiments by Turkey in what was known as "Operation Medusa". However, it must be recognised that the deal between Israel and Turkey was more important for other political and strategic concerns rather than for the enhancement of water supplies.

Instead of importing from external sources, transcatchment water transport can be initiated internally. The best example is the Great Man-made River Project in Libya which was built to irrigate some 18,000 hectares of land around the Gulf of Sirte. The water is transported some 800 km from deep aquifers at Kufra in the southern Fezzan.

Apart from overt imports, the arid belt states of the Middle East have enhanced their water budgets by covert means. By importing commodities, particularly agricultural produce, which are water-intensive to produce, they are adding what has become known as "virtual water" to their supplies. The focus of their own water use can then be crops and other products which yield a high economic return. It was estimated that in 2000, the states of the Middle East imported 25 per cent of their requirements as virtual water. However, while this procedure may reduce fears about the sustainability of water supplies, it raises another key problem: food security.

Worldwide, food security remains a vital issue but in the Middle East it received particular prominence following the major oil price rises of 1973/74. Given the fact at that time that most of the oil producers imported at least 90 per cent of their food, consideration was aired by Western diplomats of combating the "oil weapon" with the "food weapon". Since then, food security has become an obsession particularly among the oil-rich states of the Gulf and has led to some outlandish expenditures on crop production. For example, Saudi Arabia became for a while a significant exporter of wheat, that wheat having been produced at something approaching five times the normal costs. However, food is strategic predominantly as the result of maldistribution and, above all, lack of purchasing power. Unlike water, food is virtually infinitely substitutable and it is only the poorer Middle Eastern states such as Sudan, Mauritania and Western Sahara which are likely to encounter real food security problems in which the average per capita intake of calories is deficient. At the present time, in the Middle East, only Sudan and Afghanistan are listed as states where on average people receive less than 90 per cent of their daily needs for calories per capita.

A further source of water supply, particularly in the driest areas where it can make a significant difference, is the harvesting of dew and mist. This procedure has long been practised by pastoralists who in places such as the Dhofar of Oman have built stone troughs under trees and bushes to catch drips and flows from the branches and leaves. Vegetation and animals, certainly in some parts of the Arabian Peninsula, depend upon dew for survival.

The demand side of the water budget can be addressed through conservation. Since, throughout the region, agriculture consumes about 70 per cent of the water available, the selection of irrigation techniques and of crops can make a major contribution to water savings. For example, computer-controlled drip irrigation as practised in Israel can target exact amounts of water to individual plants. However, as water is either heavily subsidised or free throughout the region, there would appear to be limited incentives to conserve it. In addition, in many of the countries there is no monitoring of water abstraction and water law enforcement is negligible. Therefore, allocation of water among the demand sectors is extremely difficult. However, with improvements in agricultural techniques, conservation has become a reality, although it is restricted to the more developed parts of the region. In the countries such as Sudan where the agricultural sector remains predominant, water use is still vast and conservation procedures are very limited.

In the Middle East, the problems connected with water are so fundamental that, in attempting to discern the most effective approach to the many issues raised, three schools of thought can be identified. There are the liberal-functionalists who point out that, despite forebodings by politicians and experts, water wars have not occurred. This is the optimistic view, exemplified by basins such as that of the Indus where water can act as a unifier for peace. Water is so crucial to every side that accommodation must be made in a cooperative manner. For the economist, water pricing is the basis of addressing the problem. Water is seen as a "good" and as a result it is possible to calculate the costs of production and to implement economies of use. The most visceral concern is the availability of potable water and only 5 per cent of the water budget is normally used for drinking. The third approach, and the one to gain most publicity, is the geopolitical in which water is crucial in political decision-making at any scale from the national to the local. An example frequently cited is the reluctance of Israel to trade West Bank land for peace as the vital West Bank aquifers would then be lost. In fact, the shortfall could be made good by desalination relatively easily but, in what is still essentially an agricultural land in its thought processes, water is thought too vital to be traded.

Since water is a universal necessity and has a greater range of possible uses than any other resource, the potential problems for sharing it are numerous. International river basins are those shared by two or more states and it is at the national level in such basins that conflict over water is considered most likely. Indeed, there has been so much speculation about such events that the term "hydropolitics" has emerged. In Africa, the number of such international continental basins is fifty-seven and in Asia, forty, the area covered being respectively 60 per cent and 65 per cent of each continent. Hydropolitically, the most extreme tensions are likely to arise in a small basin shared by several states, a basin such as that of the River Jordan. This covers only 11,500 km^2, the percentage of the area of each riparian state within the basin being Jordan 54, Syria 29.5, Israel 10.5 and Lebanon 6.

A basic problem in developing cooperative management of such basins is that of sovereignty, over which there is still dispute in international law. With the realisation that water is a vital resource which, unlike petroleum, moves across international boundaries legal concern has grown over the questions of both quality and quantity. There are four main traditional alternative principles governing the sovereignty of states over water resources. The most recent of these, and the one which would seem to offer the fairest outcome, is the principle of equitable utilisation or limited territorial sovereignty. A state may use international waters in so far as it does not interfere with their use by co-riparians. Using this principle, the International Law Association developed the "Helsinki Rules" which, with more recent refinements, are the current basis for negotiations. Where there is the potential for conflict, the major factors are:

i the relative geographical positions of the states;
ii the degree of interest in water problems demonstrated by each state; and
iii the power from internal or external sources which each state can bring to bear to influence decisions.

Other factors include disagreements over data, differing views over development and ideological differences. The importance of geographical position depends upon whether the states are contiguous or successive. The Tigris forms part of the boundary between Syria and Turkey which are therefore contiguous. The Euphrates flows from Turkey through Syria and then into Iraq and these states are therefore characterised as successive. In the Jordan catchment, there are both successive and contiguous relationships.

In the Middle East, there are several international basins but three in particular involve areas of scarcity and have given rise to hydropolitical speculation. All three have an essentially non-cooperative setting within this environmental imbalance and in all there have been instances of strong disagreement if not actual conflict. In the case of the Tigris and the Euphrates, considered to share one basin, there is theoretically sufficient water but there are major imbalances in its utilisation. In the Nile Basin deficits are already reported while for the Jordan basin most analysts consider that there is a zero-sum situation.

The Nile and its tributaries drain a basin which covers 10 per cent of Africa, an area with more distinct climatic types than any other river catchment (Map 22). The White Nile rises in the equatorial climatic zone of the African Lake Plateau while the key source, the Blue Nile, flows from Lake Tana in the monsoon climate of the Ethiopian Highlands. The discharge varies from 78–85 bcm annually and the international nature of the basin is shown by the fact that the water is shared by nine states. However, the downstream state, Egypt, is the most powerful and has by far the largest population of any of the countries involved. Egypt contributes no inputs into the Nile system but abstracts most water from it. In 1958 there was confrontation over the construction of the Aswan High Dam but agreement was reached in 1959 for the sharing of the water of the resulting lake, Lake Nasser (22 bcm) in a 2:1 ratio between Sudan and Egypt. There have been many suggestions about the construction of canals to improve discharge but the only one of real importance, the Jonglei Canal, has still to be completed. It is calculated that it would add an additional 4.7 bcm to the overall discharge.

Ethiopia contributes 84 per cent of the total Nile flow but withdraws only modest amounts. It has been unable to capitalise upon its geographical position as a result of physical geography and political weakness and has indeed been threatened by Egypt should major irrigation schemes be implemented in the Upper Blue Nile area. However, since the 1980s economic crises and civil war in both Ethiopia and Sudan have hindered development and serious competition for water is in abeyance. In fact, Sudan has tended in recent years not to use all the water available to it under the 1959 agreement. Ethiopia's most ambitious possible demand has been calculated as between 4 and 5 bcm annually, approximately 6 per cent of the total flow. In the meantime, Egypt has become a country less dependent upon agriculture. Meanwhile, the East African states in the Basin have threatened to abrogate a treaty signed in 1929 between Egypt and Britain, as the colonial power. The treaty forbids any country south of Egypt from reducing the flow of the Nile without the consent of Egypt.

The combined flow of the Tigris and Euphrates together approximates to that of the Nile (Map 23). The Euphrates has three riparian states and discharges annually 35 bcm while the Tigris has two riparians and an annual flow that varies between 49 and 53 bcm. In 1974, during the filling of Lake Assad, there was a standoff between Syria and Iraq but the main focus of potential conflict within the Basin has been the South-east Anatolian Project (GAP) which dominates water in the catchment. On 13 July 1990, the diversion channel beneath the Ataturk Dam was closed, a normal procedure in such constructions, and the Euphrates was effectively cut off for one month. Despite the fact that there had been earlier compensating flow and both downstream states had been warned, the alarm was immediately raised over water security. The psychological impact was sufficiently profound for the governments of Iraq and Syria to work closely together over the issue. At the end of the month, flow was resumed at an agreed 500 m^3 per second.

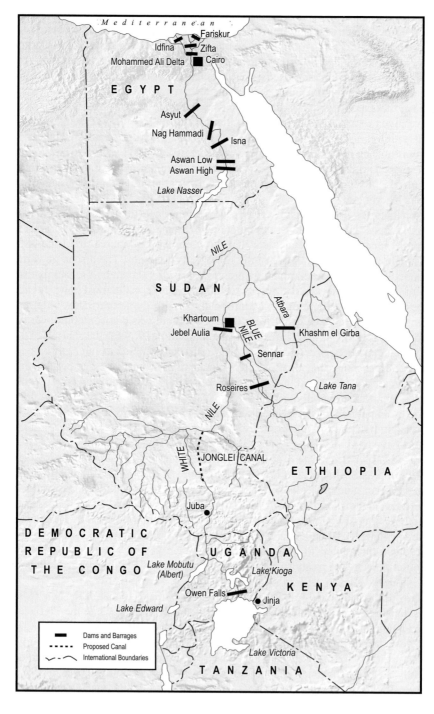

Map 22 Water management: Nile Basin

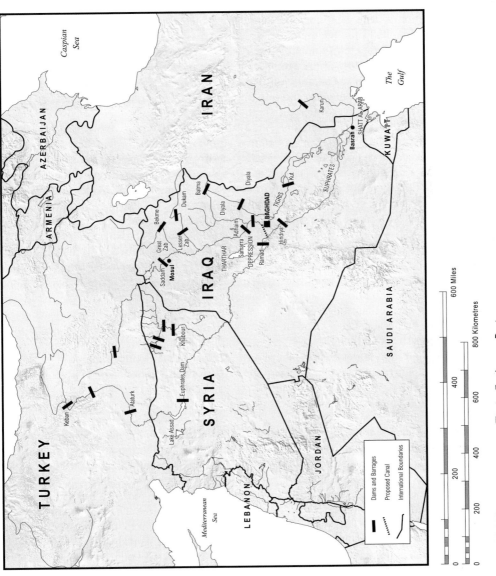

Map 23 Water management: Tigris–Euphrates Basin

Turkey's unilateral development of the upper basin has, since 1970, reduced the annual flow of the Euphrates at the Turkey/Syria boundary by almost half. The completed GAP project has 80 dams, 66 hydroelectric power stations and 68 irrigation systems. Turkey maintains that it has the right to develop water resources that originate within its territory and also that the result of its development offers the advantage of a regulated water supply to the downstream states. In order to maintain food self-sufficiency, Syria has major irrigation plans which are obviously affected by the decline in flow of both the Euphrates and the Tigris. In Iraq, 50 per cent of agricultural production is dependent upon irrigation but, isolated from its neighbours and in some political chaos, Iraq had little negotiating capability. In 1980 Turkey and Iraq established the Joint Technical Committee for Regional Waters, which Syria joined in 1983. However, tensions arose until the signing in 1987 of the Protocol on Economic Cooperation. In April 1990, Syria and Iraq agreed that the water transiting the Turkey–Syria boundary should be divided proportionately between Syria (42 per cent) and Iraq (58 per cent). In 2002, the government of Turkey agreed to open discussions over water with both Syria and Iraq.

The Jordan Basin (Map 24) covers only 11,500 km² and is therefore of an entirely different order in size from the other two. The annual discharge is 1.5 bcm, approximately 2 per cent of that of the Nile at Cairo. There are three headwater streams: the Hasbani, the Banias and the Dan, the last of which alone rises in Israel. Various seasonal wadis provide ephemeral flow into the Jordan but the only major perennial inflow is from the Yarmuk River. Water in the Basin is extremely scarce and the co-riparians: Syria, Jordan, Israel and the Palestinians; all face water shortages. With regard to the rivers themselves the situation is complex. The Yarmuk River provides the boundary successively between Jordan and Syria and Jordan and Israel. The river Jordan is the boundary between Israel and Jordan and, to the south, between the West Bank and Jordan. There have been several incidents in which water probably played a part, the most significant occurring in the early construction of the National Water Carrier by Israel. Then, following the 1964 Arab Summit a plan was put forward to divert the headwaters of the Jordan. Military strikes into Jordan followed and these continued until the 1967 War. Following that war there were further exchanges and in 1969 Israel attacked the East Ghor Canal, the main irrigation system of Jordan.

Since 1939 and the Ionides Survey, there have been at least seventeen major plans for the integrated development of the basin and the peaceful use of water. The Johnston or Unified Plan put forward in 1955 proposed an allocation of: Jordan 52 per cent, Israel 36 per cent, Syria 9 per cent and Lebanon 3 per cent. The plan was generally accepted although not ratified but it has remained as a guideline.

In examining the possibilities for conflict, three key locations within the Basin of the Jordan can be identified. The Yarmuk Triangle allows Israel to influence the apportionment of the Yarmuk water and thereby affect the inflow to the East Ghor Canal. Occupation of the Golan Heights provides a little additional water for Israel but, more importantly, allows control of the headwaters of the Jordan. Occupation of the West Bank gives access to the three main aquifers: the Eastern Aquifer, the North-east Aquifer and the Western Aquifer. The total available water from the three aquifers is on average 679 mcm/y, a figure equivalent to 45 per cent of the flow of the Jordan. An agreement in 1995 between Israel and the Palestinians shared the water respectively as follows: 483 mcm/y to 196 mcm/y. The figure for Israel is similar to that produced by desalination and could be covered relatively easily by an increase in the desalination programme. The other hydropolitical element concerns plans for a possible diversion of the Litani River in southern Lebanon. Since the withdrawal of Israel from the region this appears unlikely but it is known that a survey was carried out which revealed that the engineering problems would be immense.

Map 24 Water management: Jordan Basin

International boundaries

At the start of the twentieth century, Lord Curzon was moved to state that frontiers were the razor's edge on which hung suspended the great issues of war and peace. A century later, after two world wars and a prolonged Cold War, boundaries remain a crucial concern of international relations although there has been a change in emphasis. As economic interdependence grows through globalisation, there is if anything an even greater requirement for a stable territorial order as only state power can enforce the rules.

The state is an internationally recognised political and juridical entity claiming sovereignty over a specific area of land and possibly adjacent sea, the inhabitants of the area and the resources located therein. Ideally, and in most cases, this area is delimited by boundaries shown on maps and in some cases demarcated on the ground. Boundaries identify the area to which the nation owes allegiance and the extent of government control. They demonstrate the territorial integrity of the state which is, in the majority of cases, legally recognised by neighbouring states and the international community. However, since 1989 there have been major changes in the pattern of states so that at the present time approximately one-quarter of the land boundaries can be classified as unstable. Furthermore, almost two-thirds of the potential global maritime boundaries have yet to be settled. Since the possession of agreed boundaries is vital for the operation and security of the state, conflicts related to boundaries are likely to remain a significant element in international relations. It is not an overstatement to say that the functioning of the global order depends upon territorial integrity and therefore the settlement of boundary issues.

The concept of territoriality presupposes the existence of boundaries. The functioning of the state as an entity depends upon boundaries. Elements of statehood ranging from the resource base to national law, the census, elections, security and iconography are dependent upon the possession of clearly defined territorial boundaries. The link between territory and national sovereignty has been clearly demonstrated by the Palestinians who were willing to accept even the minimum area of the Gaza Strip and the environs of Jericho as a basis for their state. Immediately after the Second World War, the Kurds made every effort to develop what was only a sliver of their potential area in the state of Mahabad. A nation without a defined territory has very little standing in the world order.

In the world at large, the parallel processes of fission and fusion have been seen, for example, in the cases of the Soviet Union and Vietnam, respectively. In rather less dramatic circumstances in the Middle East, the recognition of Palestinian entities represents fission while in Yemen the unification of the Yemen Arab Republic (YAR) with the People's Democratic Republic of Yemen (PDRY) provides a good example of fusion. With the settlement of the maritime boundary between Qatar and Bahrain by the International Court of Justice (ICJ), the new maritime boundary is an example of fission but within fusion since both states belong to the Gulf Cooperation Council (GCC).

Boundaries can be classified in numerous ways depending upon the field of interest. For trans-boundary movement, the continuum from complete cut-off or isolation to total integration can be used. In general, the boundaries of Israel and the Occupied Territories would be considered isolated, particularly with the development of the Israeli Separation Barrier. In many aspects, the boundaries of the separate Emirates are integrated within the UAE. Related to this is the degree of boundary porosity, either allowed or perceived. For example, the boundary between Saudi Arabia and Jordan, settled in 1965, allows for a penetration to a depth of 20 km either side for pastoralists using traditional grazing and water rights. Throughout the core of the Middle East, where there are few relief features, boundaries are porous partly at least because there is nothing to mark their presence. One result of this has been that increasingly states have demarcated their boundaries by pillars which are intervisible. In the more extreme cases, ditches have been dug and fences erected. A fence separates Saudi Arabia from Iraq while a new barrier has been erected between Buraimi and Al-Ain to control movement between the UAE and Oman.

Boundaries may also be classified according to whether delimitation was before or after the settlement and development of the area. Superimposed boundaries are those constructed on a developed landscape, for example those imposed by the colonial powers on the states of the Levant. In these cases, there needs to be political, social and economic adjustment to the boundaries. In the case of Palestine, Jericho presents an example of a superimposed boundary while the Gaza Strip is an example of boundary conversion. Having been the boundary between Egypt and Israel, it now delimits one part of the potential state of Palestine.

Boundaries throughout the arid core of the Middle East were mainly imposed by colonial powers and, in the absence of development or physical features, tend to be straight lines, arcs or other geometrical constructs. These are known as antecedent boundaries and, since they were drawn to recognise the extent of colonial powers rather than the local geography and geology, they can give rise to disputes particularly over oil and water resources. For example, the northern boundary between Iraq and Kuwait, marked originally by a notice board and palm trees, cuts across the mega-oilfield of Rumaila. Disputes over the extraction from this field by the two neighbours continues today. By definition, antecedent boundaries were constructed largely in ignorance of underlying resources.

There are also relict boundaries, indicating where a former boundary existed. The human landscape adjusts to a boundary and therefore when it is removed those adjustments provide evidence of the line of the former division. Such a line is still seen at least in part of the landscape that separated the YAR and PDRY.

A more general classification of boundaries, and indeed one used by Lord Curzon, was to distinguish between "natural" and "artificial". Natural were those in which the line could follow some feature in the landscape in contrast to artificial boundaries which may cut across features or may be used where there are no obvious features. This is, of course, a false distinction in that all boundaries are artificial but there are advantages with so-called natural boundaries. They indicate to the inhabitants of the state and their neighbours exactly where the change of jurisdiction occurs. It is then very clear if a boundary is transgressed. There is also a psychological element to the argument in that it is possible to see literally where a territory ends. If the boundary is a line, even a demarcated line, across a featureless plain, there is little to distinguish "them" from "us".

An amplification of this classification, and a set of categories most frequently used, is:

i   physiographic
ii geometric
iii anthropomorphic, and
iv compound.

Physiographic boundaries are those formally designated natural and comprise features of the landscape ranging from rivers, watersheds and coastlines to marshes and forests. The most obvious limitation on the state is of course the coastline but there may be offshore islands and, with the advent of maritime claims for a 12 nml wide territorial sea, possibly a 200 nml exclusive economic zone and even beyond that the edge of the continental shelf, this boundary is no longer as clear as it once was. Linear features such as rivers fulfil psychological needs and provide a line on a map but they divide drainage basins which tend to have a natural unity, particularly with regard to water supplies and agriculture. Deserts, marshes and forests are all areas rather than lines. They all have the advantage of inhibiting movement but offer little guidance for delimitation. In the Middle East, wadis may offer substitutes for rivers as boundary lines but they, like rivers, can readily change position. There are mountain ranges in the Middle East but most are located well within the territory of states and there are few watershed boundaries.

Geometric boundaries are commonly based upon latitude and longitude or straight lines linking known boundary positions. In the case of Kuwait, there were other geometrical constructions. Throughout the arid part of the Middle East, geometric boundaries predominate and, particularly in the Arabian Peninsula, the delimitations have been enhanced by demarcation.

Anthropomorphic boundaries follow recognised tribal and other traditional territorial divisions and these have been relevant in certain key Middle Eastern delimitations. The eastward continuation of the demarcated Treaty of Taif boundary between Saudi Arabia and Yemen was defined by tribal occupance.

Compound boundaries comprise any combination of the other three categories. Kuwait displays these various boundary categories. The western boundary of Kuwait follows Wadi al-Batin which marks the division of territory between tribes and is also a physiographic feature. Rather than follow the line of the deepest part of the wadi (*thalweg*), which would produce an incredibly distorted boundary, the line has been simplified and therefore has some elements of a geometric boundary. The northern boundary with Iraq is latitudinal and is continued by a straight line to the coast. The maritime delimitation follows the Low Water (Springs) of Khor Zubari and then a median line down the Khor Abdulla between Warbah Island and the mainland of Iraq.

Under International Law, since it has the right of exclusive sovereign jurisdiction over territory, control of the state is essentially territorial. In sharp contrast, according to Islamic Constitutional Law a different model for the state can be defined. Since the fundamental purpose of the government is to defend and protect the faith rather than the state, the basis of the Islamic state has been ideological rather than territorial. Furthermore, the entire Islamic world is envisaged as a single unit, the *Umma*, within which the precepts of *Shari'a*, Islamic Law, could be fulfilled. Thus, the Islamic state has been concerned with community rather than territory and there could be no question of sovereignty over unoccupied land. Allegiance was owed to a person rather than a piece of land. Therefore, while over time the Western model of political units has been accepted as providing a structure for Islamic society, there was virtually no concern over boundaries until the oil era. It is possible to plot on a map the basic location of the different tribes but there is rarely a clear-cut boundary between them and tribes are not necessarily static.

The most important influences on territory and boundaries were the Ottomans before the First World War and France and Great Britain afterwards. Following the Sykes–Picot Agreement (1916) and subsequent treaties, France and Great Britain redrew the political map, producing boundaries which often cut across existing social and economic divisions as new states were created. This left the legitimacy of the boundaries and the states enclosed by them open to question as the divisions were imposed largely for the convenience of the delimiting powers. Since the end of the Cold War, the world political map has become more malleable and there is evidence of this in the Middle East today. In 1991, the two Yemens united and in 1995 Palestine finally emerged as an independent entity. In the Asian part of the Middle East, the colonial boundaries remain but their acceptance is dependent upon diplomatic and legal procedures ranging from negotiation to arbitration. In the North African part of the Middle East, the concept of *uti possidetis* was accepted in the Cairo Declaration of the Organisation of African Unity (OAU) in 1964. This meant that the states agreed to retain the colonial boundaries, despite their obvious limitations, to avert ensuing chaos. *Uti possidetis* underpinned the land boundary settlement by the ICJ in the case of Libya and Chad.

For the allocation and delimitation of boundaries, the Middle East presents a number of problems. In most Middle Eastern countries, the population distribution is extremely uneven with major concentrations either along the coast or inland at sources of water. The result is that large areas of most of the countries are either uninhabited or inhabited temporarily by nomadic people. Therefore, for much of their length boundaries may appear to have little relationship to the human occupance of the landscape and there are likely to be problems of boundary control.

The Middle East is also characterised by the importance of subterranean resources, particularly water, oil and natural gas. While international boundaries on the surface may be meticulously delimited, the boundaries of the resource fields beneath the ground are unlikely to be exactly located as their full extent may well be unknown. Thus, there is likely to exist a mismatch between the limits of sovereignty and those of crucial underground resources.

There is no real prescription for land boundary settlement which may depend upon geology, geography, anthropology or political, military and security concerns. Whether in face-to-face meetings between state representatives or through the ICJ there is a process of negotiation unless an external mediator is employed. Maritime boundaries present a rather different range of problems in that there is a menu. The UN Conference on the Law of the Sea (UNCLOS) which came into force one year after ratification by the sixtieth state in November 1994, provides general guidelines for defining boundaries. In the Middle East the major difficulties include the fact that the seas tend to be enclosed or semi-enclosed with very few countries having access to a full 200 nml exclusive economic zone (EEZ). In fact, the full EEZ can be claimed by only Oman and Yemen. Added complications in the Middle East are the occurrence of strategic islands and the offshore location of many oil and natural gas fields.

That said, over the recent past the Middle East has been the leading region for boundary settlement (Map 25). Land boundaries settled through the ICJ include Egypt and Israel over Taba (1988) and Libya versus Chad over the Aozou Strip (1994). The Kuwait/Iraq boundary was settled, delimited and demarcated by the UN (1993–4). Boundaries settled relatively recently by negotiation include Saudi Arabia/Oman (1990), Saudi Arabia/Qatar (1993) and Saudi Arabia/Yemen (2000).

The maritime settlement between Saudi Arabia and Bahrain, which included accommodation for the Fasht Abu-Sa'fah oilfield (1958), was the first maritime settlement in the Gulf. Settlements in the Middle East have been achieved through the ICJ between Libya and Malta (1985), Libya and Tunisia (1988) and Bahrain and Qatar (2001). Saudi Arabia and Iran agreed their maritime boundaries in 1968 and Saudi Arabia negotiated boundaries with Kuwait and Yemen in 2000. Yemen agreed the maritime boundary with Eritrea in 1999.

Nonetheless, there are several outstanding boundary issues in the region, some of which have resulted in conflict (Map 26). The Aegean Sea maritime boundary between Turkey and Greece, given the disposition and sovereignty of the various island groups, presents almost intractable problems. To negotiate navigation, fishing and seabed extraction rights would probably require a set-aside of sovereignty, a procedure successfully used in the case of the Falkland Islands. The division in Cyprus between the Turkish and the Greek populations is long-standing but appears to be nearing some sort of settlement. The most difficult issue of all, of course, concerns Israel and the Palestinians. Eventually it is assumed that the new state

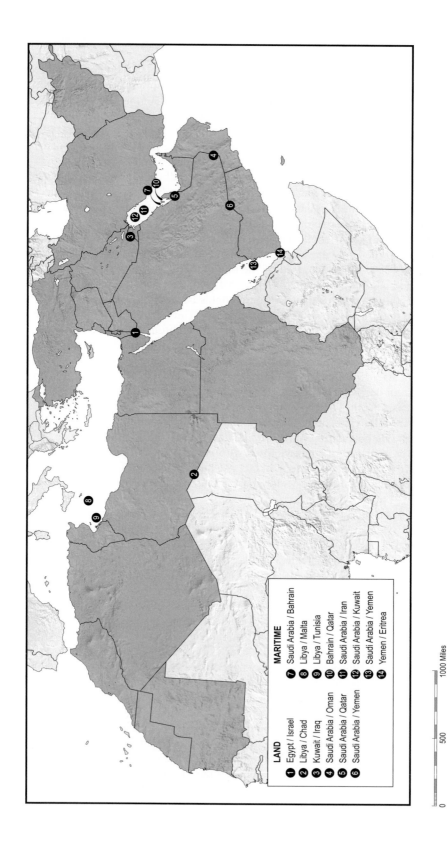

1000 Miles

1500 Kilometres

Map 25 Recently settled boundary disputes

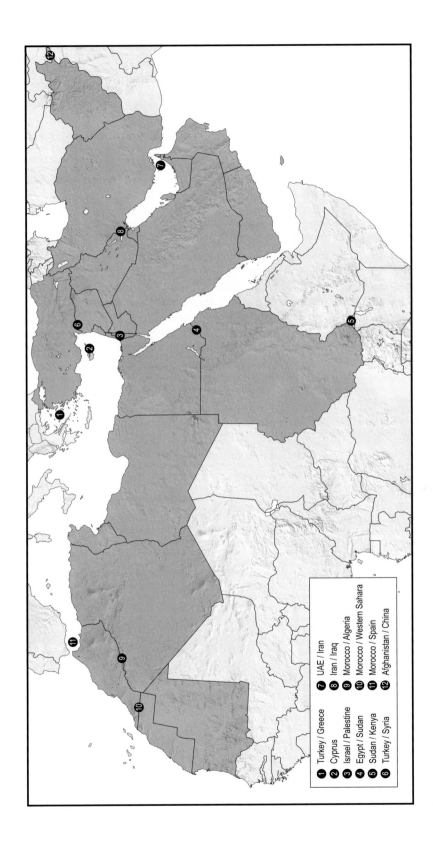

Map 26 Key boundary issues

1 Turkey / Greece
2 Cyprus
3 Israel / Palestine
4 Egypt / Sudan
5 Sudan / Kenya
6 Turkey / Syria
7 UAE / Iran
8 Iran / Iraq
9 Morocco / Algeria
10 Morocco / Western Sahara
11 Morocco / Spain
12 Afghanistan / China

0 500 1000 Miles
0 500 1000 1500 Kilometres

of Palestine will see the remainder of the West Bank added to the Gaza Strip and the little area of Jericho which presently represents the potential state. A key issue will be the proportion of the West Bank which is added. Egypt has a long-running boundary problem with Sudan over the Halaib region in the south-east and, in a parallel location, Sudan and Kenya have yet to agree on the Ilemi Triangle. A similar situation has obtained long-term between Turkey and Syria over Hatay which remains claimed by Syria. Hatay also includes the lower regions of the Orontes River and therefore presents a potential hydropolitical issue.

The long-running maritime dispute continues between Iran and the UAE over the islands in the approaches to the Strait of Hormuz: Abu Musa and the Tunbs. After the conclusion of the Gulf War (1991) it would appear that the boundary dispute between Iran and Iraq over the Shatt al Arab has been settled but this particular issue has presented a long-term recurrent problem. Morocco still has land boundary settlements to be agreed with Algeria and the Western Sahara. There are also long-running disputes with Spain over the enclaves, primarily Ceuta and Melilla. In Afghanistan, the eastern end of the Wakhan Panhandle is still disputed by China.

Apart from these issues, all with a history, maritime boundary settlements still have to be made between Kuwait, Iran and Iraq at the head of the Gulf, throughout most of the Red Sea, where the only settlements have been between Saudi Arabia and Yemen and Yemen and Eritrea, and at the southern end of the Caspian Sea where Iran has been unable to reach agreement with the other riparians. In the Mediterranean, the only Middle Eastern state to have negotiated all of its boundaries, other than the adjacent boundary with Algeria, is Tunisia. Libya has settlements through the ICJ with Tunisia and Malta and the remaining states have still to pursue their claims.

Transboundary issues

One element of state sovereignty is the control of entry and exit across boundaries. In most states, control is conspicuous at authorised crossing points on the land boundary, at specific sea ports on the coast and at international airports. Elsewhere along the land boundary there may be patrols and, increasingly commonly, fences or various forms of detection. Depending upon the physical landscape, transboundary movement and detection range widely in difficulty. A geometrical boundary across a desert plain may be far easier to cross than a mountain range but detection is far simpler. Therefore, there are natural controls which render a boundary more or less porous. On the coast, between authorised entry ports, there are likely to be smaller harbours and inlets together with isolated uninhabited stretches. In the Middle East the coast guard system, whether operated from land or sea, is of limited effectiveness. It must also be borne in mind that the legal limits of the state are beyond the coastline and are represented by claims for territorial waters and exclusive economic zones or agreed boundaries with neighbouring states. To control the waters of a state requires relatively sophisticated naval patrols and few Middle Eastern states have significant navies. For well-organised groups, illegal entry may be effected through small airfields or unmetalled runways. The desert landscape is particularly suitable for the operation of illegal airstrips.

The extremes of government control can be seen in the cases of Israel and Afghanistan. Israel has a highly effective boundary patrol system enhanced by fences and, in the case of the West Bank, the Separation Barrier. The coastline is patrolled by a small but effective navy. The limited size of the country means that surveillance is considerably less difficult than in the cases of the larger Middle Eastern states. In contrast, Afghanistan is a large state with mountain and desert boundaries. Furthermore, only one of the major constituent tribes is located wholly within the country. In all other cases the boundaries dissect the territory of tribal groups with, in most areas, little obvious control. Boundaries of Afghanistan are known for their porosity, a key factor in the current conflict.

Apart from problems specific to boundaries and entry points, there is also the question of the effectiveness of customs. This depends not only upon the integrity of the customs officers but their level of training and their experience. Above all, the customs service must be supported by appropriate government legislation and regulations which are clearly enforced. For each state of the Middle East, the main shortcomings with regard to drug smuggling, people trafficking and money laundering are described appropriately in Section D. A perusal of these illustrates the point that illegal movements are transboundary not only between countries of the Middle East but from and into the Middle East as a region. Some, such as arms trading and arms smuggling or immigration and people trafficking, illustrate gradations from the legal to the illegal. The key issues are developed further in Section E.

Illegal transboundary movement occurs throughout the world for a variety of reasons. In the Middle East, apart from the universal factors ranging from the quest for a better life through to ideology or greed, there are specific factors which can be related to illegal movement. Most of the states have authoritarian regimes and the resulting suppression of freedom in many cases gives rise to tensions and has triggered the formation of deviant and extremist groups. In many, perhaps half of those generally identified as terrorists, there is strong ideology resulting from the particular perceptions of Islam. The hatred engendered appears to be mostly directed against the West and, if threats are put into practice, requires transboundary movement. Opposition to the regime and internal conflict may lead to flows of political refugees. The Middle East is also known for the movement of economic refugees.

The Middle East includes countries with some of the highest per capita gross domestic products in the world and these oil-rich states act as a magnet for labour from the Indian subcontinent and the poorer countries of south-east Asia. The division between rich and poor among the states of the Middle East also, of course, results in flows across borders within the region. However, dependence upon oil is not wholly beneficial in that the volatility of the market can result in massive and sudden upturns and downturns in the oil-rich economies. These affect the migrations of the population. There are also significant social problems within the region, particularly youth pressure. In general, the populations of the Middle East demonstrate rapid growth with the result that some 45 per cent of the population is under 18 years of age. This results in pressure on all the services, particularly education, and also in underemployment and unemployment. All of these can generate deviance, through crime or terrorist activities, and the movement of people. A further notable feature of the Middle East has been the rate of urbanisation which has produced vast sprawling cities with little in the way of infrastructure and services. Unemployment, urban squalor and ghetto-living can all generate unrest in some form and result not only in conflict but also in illegal activity and movement. Thus, in many parts of the region, social, economic and political failure have all contributed to factors that can give rise to transboundary movement (Map 27).

In considering these factors, it must be borne in mind that the developed world and particularly the West have given rise to some and at least contributed to many of the problems. At some stage the global need for oil together with the power of governments and multinational corporations and the actions of the global oil market have all contributed to the problems associated with petroleum. Low-cost labour is encouraged not only within the oil-rich countries of the Middle East but also in the West where, for example, Turkish workers have been encouraged but have then become the subject of discrimination. The narcotics trade is demand-led and it is the demand of the West in particular which has produced the transboundary drugs movement of the Middle East.

Of all the illegal transboundary movements, it is terrorism which sparks the most fear, narcotics which are the most insidious and people trafficking or modern slavery which inspires the most horror. More terrorist groups have been identified in the Middle East than in any other region and this has given rise to a range of half-truths and myths about Islam and fundamentalism whereas, in fact, half at least of the groups are secular. Given the variety and disposition of the groups, it is often difficult to interpret whether the terrorism was transboundary or indigenous. However, it can be said with some certainty that the driving force of many Middle Eastern terrorist groups is hatred of Western values and those associated with them. Targets may therefore range from government officials to foreign military personnel, tourists or state governments aligned with the West. Further impetus is given to extremism by US power projection into the region and by Western support for Israel in its long-running conflict with the Arab world. While in some states, for example Iran, there is support for terrorist groups such as Hezbollah, actual state-sponsored terrorism is more difficult to verify. It depends very much upon definition. Is an assassination in a foreign country by Mossad state-sponsored transboundary terrorism?

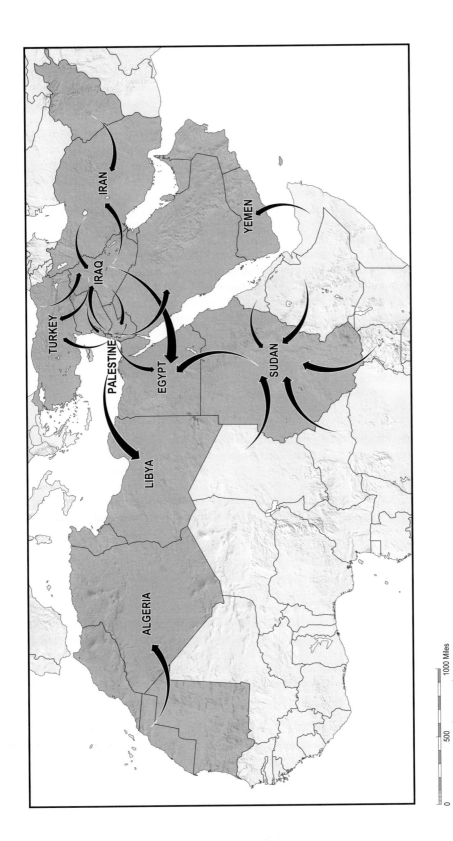

Map 27 Middle East: refugee movements

Terrorism clearly has connections with international crime and with arms smuggling. One fear is that as terrorist groups become more sophisticated and enjoy greater financial support they may venture into using weapons of mass destruction. Cyber-terrorism is already an issue and the global development of computing has clearly been used by terrorist groups to their great advantage.

The Middle East is in many ways the focus of the arms market. In the period 2003–6, arms deliveries to the Middle East were valued at $43 thousand million, of which $18.7 thousand million went to Saudi Arabia. The major suppliers were the countries of Western Europe, particularly the UK, together with the USA, Russia and China. With such huge amounts of hardware coming into the region together with major conflicts in Iraq and Afghanistan, transboundary arms transfers are rife. A major fear is that weapons of mass destruction, chemical, biological and nuclear or a cocktail of all three, may be transferred across boundaries. However, it must be borne in mind that chemical weapons are unlikely to be effective on a large scale and their use is relatively difficult. Furthermore, when Iraq deployed gas against its own people and the Iranian army, there was little obvious disapproval from the West. Nuclear weapons demand a degree of sophistication both for their assembly and use which is probably beyond most extremist groups. It is biological weapons which are easy to employ and are capable of widespread destruction, which could cause major problems.

Arms smuggling is an occupation not only of terrorists but also of organised crime. However, the chief source of income for organised crime is thought to be the transboundary transfer of illicit drugs (Map 28). It is estimated that 90 per cent of the profit goes to the traffickers and only 2 per cent to the growers. Nonetheless, for impoverished farmers, the crop may be crucial and its eradication, without the provision of substitute crops, can cause ruin. Of the Middle Eastern countries, two are outstanding producers of drugs. Afghanistan is the largest opium producer in the world and Morocco is the largest hashish producer. Apart from those producer countries, major transit and trans-shipment countries in the Middle East are Egypt, Iran, Lebanon, Syria, the UAE and especially Turkey. Lebanon is also a producer of cannabis. Money generated by drug smuggling and other illegal activity leads to money laundering; particularly significant in this regard are Afghanistan, Cyprus, Egypt, Iran, Israel and Syria. Israel is also a key importer of drugs.

Drug money also helps support the network for illegal immigration and people trafficking. Illegal immigrants, seeking a better economic future, are likely to lead a precarious life in most countries of the Middle East. Their existence will be associated with poverty and crime. Details of people trafficking are included in the descriptions of states in Section D. The Middle East is a destination for men, women and children trafficked from south and east Asia, eastern Europe, Africa and other parts of the Middle East for involuntary servitude and sexual exploitation (Map 29). In the UAE alone it is estimated that 10,000 women may be victims of sex trafficking. Women also migrate from Africa, south Asia and south-east Asia to work as domestic servants but may have their passports confiscated, be denied permission to leave the place of employment or face sexual or physical abuse by their employers. Men, particularly from south Asia, come to work in the construction and related industries but may be subjected to conditions of involuntary servitude, having to pay off travel costs and sometimes having their wages denied for months at a time. Children are still trafficked as camel jockeys but are also brought in for involuntary servitude and sexual exploitation. There is a horrifying intersection between the sex trade and forced labour resulting from the huge numbers of desperate and displaced people throughout the Third World. Those deemed to be without rights can be abused for profit and this may be viewed as part of the national economy. The Middle Eastern countries most associated with the trafficking of people for labour and sexual abuse are Algeria, Bahrain, Cyprus, Egypt, Iran, Israel, Sudan and the UAE. In both Libya and Sudan slavery is clearly recognised but it is difficult to disentangle the entire trade from slavery.

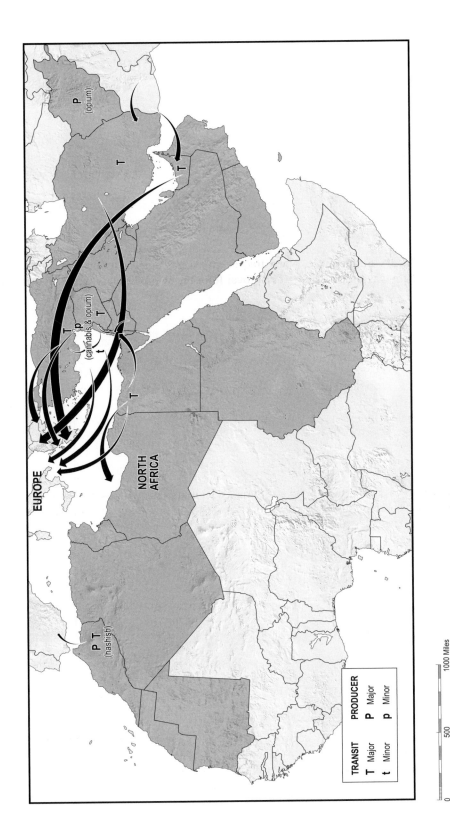

EUROPE

P
(opium)

T

T

T

(cannabis & opium)
T

P

t

T

NORTH
AFRICA

P T
(hashish)

TRANSIT   PRODUCER
T Major   P Major
t Minor   p Minor

1000 Miles
1500 Kilometres

500
1000
500

0
0

Map 28 Middle East: drug smuggling

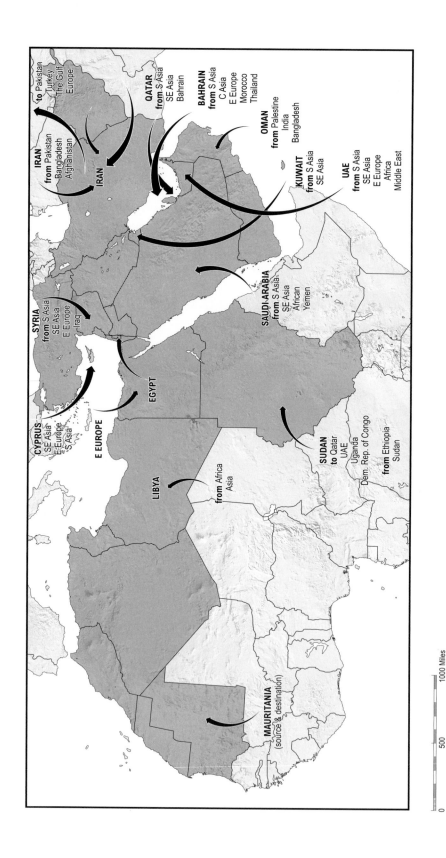

Map 29 Middle East: people trafficking

Section D

States of the Middle East

The procedure by which the description of the total Middle Eastern landscape was built up from the physical basis through the socio-economic and cultural systems to the political overlay is equally applicable to individual states. For each state, the components summarise the main elements of the physical, socio-economic, cultural and political/military systems. At the same time, the analogy with the theatre can be pursued further and the components have been refined to offer an assessment of the state's power to act on the Middle Eastern and global stage. Power in this context can be related to such factors as location, area, population, economic and military strength and external support through connectivity. It may be exemplified or enhanced through the action of an influential individual.

The tabulated components offer a synopsis of the current situation but it must be remembered this also results from history and a list of recent significant events is provided. What is at stake currently for each state is considered in a section on "Key Issues" and conclusions are reached on the current overall status of each state.

The main sources for this section have been highlighted in the Preface but are set out in more detail in Section F: Further Reading. Most of the terms used are in common parlance but a number, together with the various indices, require further definition. The majority of these definitions are synopses of the definitions which appear in *The World Factbook* and the provenance of the remainder is listed.

Boundary Vulnerability Index   The source of the Index is *International Boundaries: A Geopolitical Atlas* (Anderson 2003). This is a national assessment based upon a wide variety of sources and discussions with experts. For each separate boundary of the state, the potential geographical accessibility has been assessed on a five-point scale (one is the least accessible), taking into account: altitude, relief and the existing communication network. A similar measure for potential political instability, with one as the most stable, has been based upon the relationship: political, economic, social and particularly military; between the states, together with specific boundary zone and boundary line problems. The measure of accessibility indicates constraints to action while the measure of instability represents the likelihood of tension or conflict. The product of the two indices, each measured on a five-point scale, expresses the potential security concerns associated with that specific boundary. The national assessment is produced by weighting the indices for each boundary according to the relative length of that boundary compared with the total state land boundary length. The resulting boundary vulnerability index for the state is presented to the nearest 0.5.

Contiguous zone   This zone is contiguous to the territorial sea and over it a state may exercise the control necessary to prevent a variety of infringements. The contiguous zone may not extend beyond 24 nml from the baseline (i.e. 12 nml beyond the territorial sea).

Continental shelf   This comprises the seabed and subsoil of the submarine areas that extend beyond the territorial sea throughout the natural prolongation of the land to the outer edge of the continental margin or to a distance of 200 nml from the baseline. Where the continental margin extends beyond 200 nml from the baseline, coastal states may extend their claim to a distance not to exceed 350 nml from the baseline or 100 nml from the 2,500 m isobath (line joining places of equal depth).

Defence budgets   Some budgets include Foreign Military Assistance (FMA), the source of which is given in parenthesis.

Exclusive Economic Zone (EEZ)   A zone beyond and adjacent to the territorial sea in which a coastal state has sovereign rights for exploring and exploiting, conserving and managing the natural resources of the waters, seabed and subsoil. The outer limit of the EEZ shall not exceed 200 nml from the baseline.

Exclusive Fishing Zone   In cases in which a state has chosen not to claim an EEZ but rather to claim jurisdiction over the living resources, the term Exclusive Fishing Zone is often used although the term is not used in the UNCLOS. The breadth of this zone is normally the same as the EEZ or 200 nml.

Failed State Index   The Index (2007) is produced by the Fund for Peace and *Foreign Policy*. Twelve indicators of instability are assessed, each on a ten-point scale, and the total represents the Index and allows the state to be ranked. Sudan is ranked first with a total of 113.7 and Iraq second with a total of 111.4. The indicators are identified as: demographic pressures, refugees and displaced persons, group grievance, human flight, uneven development, economy, delegitimisation of state, public services, human rights, security apparatus, factionalised elites and external intervention.

Freedom in the World Rating   The Rating (2006) is produced by Freedom House and is calculated using political rights (PR) and Civil Liberties (CL), both measured on a one to seven scale with one representing the highest degree of freedom. The average of these two measures is taken to indicate the status. An average of 2.5 and below provides a status of "free" (F) an average of between 3.0 and 5.0 is designated "partly free" (PF) and those states with an average of between 5.5 and 7.0 are considered "not free" (NF).

Globalisation (Index of)   This Index (2007) is produced by *Foreign Policy*. The Index is derived from twelve variables which are divided into four groups: economic integration, personal contact, technological connectivity and political engagement. As a result, it is also possible to separate indices for economic globalisation, social globalisation and political globalisation.

Gross domestic product (GDP) (purchasing power parity)   This figure (2008) represents the value of all final goods and services produced within a nation in a given year. The GDP at purchasing power parity exchange rates is the sum value of all goods and services produced in the country valued at prices prevailing in the USA. It is the measure most economists prefer when comparing living conditions or the use of resources between countries. Since there is a continuous updating process for GDP in *The World Factbook*, the dates for the latest statistics do not necessarily coincide for each country. Therefore, the latest data at the time of writing are listed.

   As appropriate in the statistics, Bahrain is used as a comparator. In the case of GDP, it is felt that the comparison is best indicated in the following way: for states with a GDP larger than that of Bahrain, multiples of the Bahraini figure are used. For states with a GDP less than that of Bahrain, a percentage is used. The same procedure is adapted for area and population.

Human Development Index Rank   This Index Rank (2007) results from a complex calculation taking into account thirty-four groups of variables. These include poverty, literacy, education, technology, economic performance, trade, energy, crime and justice, refugees, human rights and gender issues.

Illicit drugs   Coverage includes five categories of illicit drugs: narcotics, stimulants, depressants (sedatives), hallucinogens and cannabis.

Independence   For most states this was the date on which sovereignty was achieved. However, the date may represent some significant nationhood event such as the date of unification in the case of Yemen.

Literacy   There is no universal definition or standard for literacy and the rates listed are based on the most common definition: the ability to read and write at a specified age.

Maritime claims   Full definitions occur in the UN Conference on the Law of the Sea (UNCLOS) which came into force in November 1994.

Martial Potency (Index of)   The Index produced by the Royal United Services Institute (RUSI) expresses a state's ability to use its Armed Services beyond its boundaries. The scale used is logarithmic and is based primarily upon defence spending and manpower statistics although it is adjusted to take into account such factors as commitment to and preparedness for military action. The figures are for 2003, the latest available from RUSI.

Nautical mile   The international nautical mile (nml) adopted by the International Hydrographic Organisation (IHO) measures 1,852 m.

Refugees and internally displaced persons (IDP)   According to the UN Convention a refugee is: a person who is outside his/her country of nationality or habitual residence; has a well-founded fear of persecution because of his/her race, religion, nationality, membership in a particular social group or political opinion; and is unable or unwilling to avail himself/herself of the protection of that country, or to return there, for fear of persecution. The UN Relief and Works Agency (UNRWA) for Palestine refugees in the Near East uses an operational definition for a Palestinian refugee: a person whose normal place of residence was Palestine during the period 1 June 1946–15 May 1948 and who lost both home and means of livelihood as a result of the 1948 conflict. Internally displaced persons are not specifically covered in the UN Convention but are those who have left their homes for reasons similar to refugees but who remain within their own national territory.

Territorial sea   Sovereignty of a coastal state extends beyond its land territory and internal waters to the adjacent area of sea together with the air above the sea and the seabed and subsoil below it. The breadth of the territorial sea may be established up to a limit not exceeding 12 nml from the baseline (the line of the coast in simplified form).

Thalweg   The line joining the points of maximum depth along a river channel. It is also likely to indicate the navigation channel and the two terms are frequently used interchangeably.

Trafficking in persons   Trafficking in persons is modern-day slavery, involving victims who are forced, defrauded or coerced into labour or sexual exploitation. Human trafficking is a multidimensional threat depriving people of their human rights and freedom, risking global health, promoting social breakdown, inhibiting development by depriving countries of their human capital, and helping fuel the growth of organised crime. Following the Trafficking Victims Protection Act (TVPA) passed by Congress in 2000, the US Department of State produces an Annual Trafficking in Persons Report which assesses the government response in some 150 countries. Countries are rated in Three Tiers, based upon government efforts to combat trafficking. The definitions are as follows:

– *Tier Two Watch List* countries do not fully comply with the minimum standards for the elimination of trafficking but are making significant efforts to do so, and meet one of the following criteria:

i   they display a high or significantly increasing number of victims;

ii   they have failed to provide evidence for increasing efforts to combat trafficking in persons; or

iii they have committed to take action over the next year.

– *Tier Three* countries neither satisfy the minimum standards for the elimination of trafficking nor demonstrate a significant effort to do so. Countries in this tier are subject to potential non-humanitarian and non-trade sanctions.

AFGHANISTAN

Location: A landlocked, buffer state between the Indian subcontinent, Central Asia and the Middle East. (Map 30)

Area: 647,500 km²   **% Middle East:** 4.0
          Comparison with Bahrain: 973.7×

Boundaries: **Land: length:** 5,529 km
             Contiguous countries: China 76 km, Uzbekistan 137 km, Turkmenistan 744 km, Iran 936 km, Tajikistan 1,206 km, Pakistan 2,430 km
             Maritime: **Coastline:** Landlocked
             Settlements: Land: China 1895 India 1895, 1898, 1934 Iran 1872, 1935 Pakistan 1905, 1919, 1921 Russia 1885, 1946

Boundary Vulnerability Index: 4.5

Geography: **Topography:** Largely mountainous especially north-east towards the Hindu Kush, with semi-arid plains and plateaux towards the south-east.
            Climate: Continental, semi-arid with hot summers and cold winters.

Annual water withdrawal (domestic/industrial/agricultural): 23.26 cu km/yr
          (2%/0%/98%)   **per capita:** 779 cu m/yr (2000)

Population: 31.9 million
            % Middle East: 6.3   **Comparison with Bahrain:** 45.6×
            Literacy: 28.1%
            Ethnic groups: Pashtun 42% Tajik 27% Hazara 9% Uzbek 9% Aimak 4%

Economy: **GDP:** $35 billion   **% Middle East:** 0.8
          Comparison with Bahrain: 1.2×   **GDP per capita:** $1,000
          Key resources: opium, precious and semi-precious gems
          Cropped land: 12.33%   **Irrigated land:** 27,200 km²

Human Development Index Rank: 174 (2007)

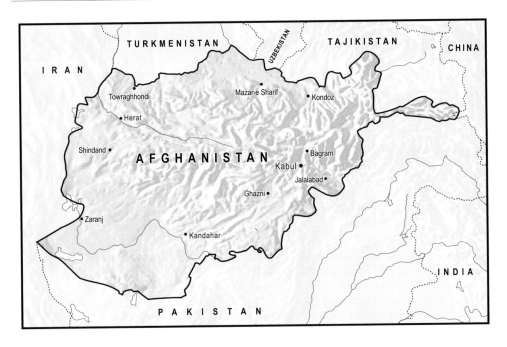

Map 30 Afghanistan

Culture: **Languages:** Afghan, Persian or Dari (official) 50%
Pashto (official) 35% Turkic 11%
Religion: Muslim 99% (Sunni 80% Shi'a 19%)
Legal system: Mixed Civil and Shari'a law

Connectivity: **Telephones (fixed):** 280,000 **Mobiles:** 2.52 million
Internet users: 535,000 **% of population:** 1.7
Imports: Pakistan 37.9% USA 12.0% Germany 7.2% India 5.1%
Exports: India 22.8% Pakistan 21.8% USA 15.2% UK 6.5%
Finland 4.4%

Index of Globalisation: − **Rank:** −

Military: Active: 50,000
Defence budget: $161 million
% Middle East: 0.2

Index of Martial Potency: − **Rank:** −

Political: **Type:** Islamic Republic   **Independence:** 19 August 1919
                       Capital: Kabul
                       Key individuals: President Hamid Karzai

Freedom in the World Rating: **PR:** 5   **CL:** 5   **Status:** PF

Failed State Index: 102.3   **Rank:** 8

Recent events

1979   Soviet invasion.
1989   Soviet withdrawal: catalyst for collapse of the Soviet Union; also by 1989 establishment of Al Qaeda by Osama bin Laden.
1996   Kabul fell to the Taliban.
1998   US air strikes against the Taliban.
2001   Post-11 September US and allied forces enter and establish an Interim Authority.
2002   US major operations against Al Qaeda.
2004   President Karzai elected.
2005   National Assembly inaugurated.
2006   Taliban insurgency begins in the south and east.

Key issues

Repatriation of refugees: 2–3 million from Pakistan.
Terrorism, and illegal activities linked to the north-west frontier area of Pakistan. Largest opium producer in the world: 80–90 per cent of the heroin for Europe.
Money laundering.
Drugs production and trafficking.
Land mines: clearance in operation since 1989.
Security and no economic recovery.

Status

As a landlocked buffer state which comprises largely mountains and semi-arid to arid plains and plateaux, Afghanistan is a very poor country. At $1,000 the per capita income is the lowest in the Middle East and it has by far the lowest literacy rate. It has the lowest Internet usage and the lowest percentage of population using the Internet of all Middle Eastern states. Apart from Mauritania, it has the lowest defence budget. Afghanistan has boundaries with six states which range from the volatile such as the western region of China to the highly volatile such as Tajikistan. The number of ethnic groups and languages illustrate the point that the population comprises largely a mixture of ethnicities from all the surrounding areas. Afghanistan has never been a recognisable state in that large areas have always been controlled by warlords, most of whom had power bases in neighbouring states. The Hazara are the only large group located wholly within Afghanistan.

Water is a problem throughout most of the lower areas of the country and agriculture has been relatively little developed other than for the production of opium. Although the country is almost 974 times larger than Bahrain, its GDP is only 1.2 times greater. In this poverty-stricken environment, the production of opium is a crucial element of the livelihood in large areas of the country. Under the Taliban, the area cultivated was drastically reduced but since US and NATO operations began in 2002 production has risen dramatically. Although Al Qaeda was largely removed in 2002, terrorism has now returned to Afghanistan and, given the lack of any economic recovery, national security is a key issue. By early 2008, the government of President Karzai was estimated to control rather less than one-third of the country. Given the continuing conflict and general lack of stability, Afghanistan is rated eighth on the Failed State Index. Afghanistan's rank of 174 out of a total of 178 states on the 2007 Human Development Index, makes the country the least developed of all Middle Eastern states.

ALGERIA

Location: Centrally located among the *Maghrib* states of North Africa on the coast of the Mediterranean. (Map 31)

Area: 2,381,740 km²   **% Middle East:** 14.5
                    Comparison with Bahrain: 3,581.6×

Boundaries: **Land: length:** 6,343 km
                    Contiguous countries: Western Sahara 42 km, Mauritania 463 km, Niger 956 km, Tunisia 965 km, Libya 982 km, Mali 1,376 km, Morocco 1,559 km
            Maritime: **Coastline:** 998 km
                    Claims: territorial sea 12 nml
                              exclusive fishing zone 32–52 km
            Settlements: Land: Libya 1911, 1956 Mali 1905, 1909, 1963 Mauritania 1909, 1963, 1985 Morocco 1845, 1972 Niger 1909, 1983 Tunisia 1901, 1970 Western Sahara 1904

Boundary Vulnerability Index: 3

Geography: **Topography:**   North to south: narrow coastal strip rising to a high plateau and the Atlas Mountains and a desert landscape.
            Climate: Warm temperature, semi-arid Mediterranean; mountain; arid desert.

Annual water withdrawal (domestic/industrial/agricultural): 6.07 cu km/yr
            (22%/13%/65%)   **per capita:** 185 cu m/yr (2000)

Population: 33.3 million
                    % Middle East: 6.6   **Comparison with Bahrain:** 47.6×
                    Literacy: 69.9%
                    Ethnic groups: Arab Berber 99%

Economy: **GDP:** $268.9 billion   **% Middle East:** 6.4
                    Comparison with Bahrain: 10.9×   **GDP per capita:** $8,100
                    Key resources: petroleum, natural gas
                    Cropped land: 3.45%   **Irrigated land:** 5,690 km²

Map 31 Algeria

Human Development Index Rank: 104

Culture:	**Languages:**	Arabic (official), French, Berber
	Religion:	Muslim (Sunni) 99%
	Legal system:	Socialist based upon French and Islamic law

Connectivity: **Telephones (fixed):** 2.841 million   **Mobiles:** 20.998 million
Internet users: 2.46 million   **% of population:** 7.4
Imports: France 22.0% Italy 8.6% China 8.5% Germany 5.9%
Spain 5.9% US 4.8% Turkey 4.5%
Exports: US 27.2% Italy 17.0% Spain 9.7% France 8.8%
Canada 8.1% Belgium 4.3%

Index of Globalisation: 45.50   **Rank:** 94

Military: **Active:**   147,000
                       Defence budget: $3.69 billion
                       % Middle East: 3.8

Index of Martial Potency: 5.58   **Rank:** 43

Political: **Type:** Republic   **Independence:** 5 July 1962
                       Capital: Algiers
                       Key individuals: President Abdulaziz Bouteflika

Freedom in the World Rating: **PR:** 6   **CL:** 5   **Status:** NF

Failed State Index: 75.9   **Rank:** 89

Recent events

1990   First free elections (Municipal and Provincial) since independence.
1991   The Islamic Salvation Front (FIS) had sweeping victories in the first round of the national elections.
1992   Army intervention before the second round of the national elections could take place.
1992–8 A continuous period of insurgency.
1999   President Abdulaziz Bouteflika placed in charge by the Army.
2000   FIS armed wing disbanded.

Key issues

Continuing armed militancy with extremist militias.
Berber autonomy.
Boundary problems with claims involving Libya and Morocco.
Western Sahara: support for the Sahrawi Arab Democratic Republic (SADR) led by President Mohamed Abdulaziz; support for Polisario Front (Popular Front for the Liberation of the Saguia el Hamra and Rio de Oro). This stance puts Algeria in direct conflict with Morocco.
Banditry in the Sahel.
People trafficking: Algeria is listed as Tier Three.

Status

Algeria is a very large country, the second largest in the Middle East, its area being marginally exceeded by that of Sudan. Among Middle Eastern states, it has the second largest number of contiguous states, a distinction it shares with Turkey and Saudi Arabia. This results in a high potential for transboundary conflict. Added to this have been the internal problems resulting from the long-running insurgency of the 1990s. The trigger for the conflict was the Algerian Army's decision to cancel elections in 1992 to prevent a probable victory by the FIS. This decision ushered in a period of intense internal violence in which up to 100,000 people are estimated to have been killed. In 1999, Abdulaziz Bouteflika won election to the presidency and implemented an amnesty for those who had fought against the state during the 1990s. As a result, large-scale violence has declined markedly. In many parts of the country, however, the situation remains fraught as a result of Islamist militancy, Sahelian banditry and unrest in the Berber population as it presses for autonomy. Algeria thus remains a flashpoint on the doorstep of the European Union (EU) with strong connections throughout Europe as indicated by the list of import and export partners.

Such political issues are underpinned by economic difficulties. The severe social and infrastructural problems hinder attempts to unite the three disparate components of the country: the narrow coastal strip, the mountainous core and the desert to the south. A related point is the heavy dependence upon one commodity, petroleum. Oil accounts for 60 per cent of the budget and 90 per cent of export earnings. Algeria has the eighth largest reserves of natural gas in the world.

The most serious external problem remains that with Morocco whose administration of Western Sahara Algeria rejects. Approximately 90,000 Western Saharan Sahraii refugees are housed in tents around Tindouf. In addition, there remains a claim to an area of south-eastern Morocco itself. Not only is there local smuggling across the boundaries but Algeria is a destination and transit point for the trafficking in persons from sub-Saharan Africa and Asia. Many of these people end in involuntary servitude and Algeria is listed as Tier Three.

BAHRAIN

Location: A group of islands between Saudi Arabia and Qatar on the southern coast of the Gulf. (Map 32)

Area: 665 km²   **% Middle East:** 0.004
              Comparison with Bahrain: –

Boundaries: **Land: length:** –
                        Only land boundary: check point on King Fahd Causeway
              Contiguous countries: –
              Maritime: **Coastline:** 161 km
                              Claims: territorial sea 12 nml
                                    continental shelf to be determined
                                    contiguous zone 24 nml
              Settlements: Anglo-Turkish Convention: 1913
                              Maritime: Iran 1971 Saudi Arabia 1958 Qatar 2001

Boundary Vulnerability Index: –

Geography: **Topography:**   Low-lying coastal fringe islands.
              Climate: Hot and arid, modified by the maritime location.

Annual water withdrawal (domestic/industrial/agricultural): 0.3 cu km/yr
              (40%/3%/ 57%)   **per capita:** 411 cu m/yr (2000)

Population: 708,600 (non-nationals 235,100)
              % Middle East: 0.1   **Comparison with Bahrain:** –
              Literacy: 86.5 %
              Ethnic groups: Bahraini 62.4% non-Bahraini 37.6%

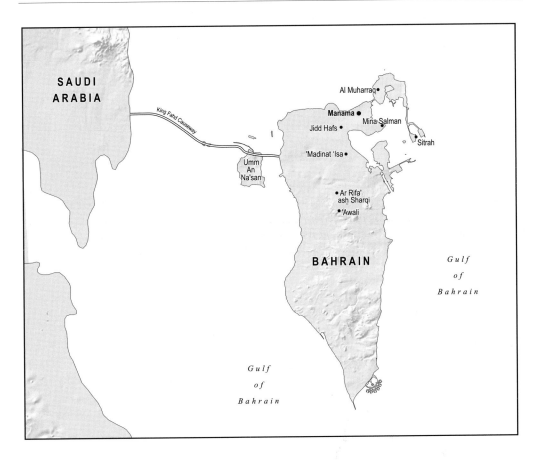

Map 32 Bahrain

Economy:

GDP: $24.61 billion
Comparison: –
Key resources: oil, natural gas
Cropped land: 8.45%

% Middle East: 0.6
GDP per capita: $35,700

Irrigated land: 40 km²

Human Development Index Rank: 41

Culture:

Languages: Arabic, English, Farsi, Urdu
Religion: Muslim 81.2% Christian 9%
Legal system: Based on Islamic and English common law

Connectivity:

Telephones (fixed): 193,300
Internet users: 157,300

Mobiles: 898,900
% of population: 22.2

Imports:	Saudi Arabia 37.2% Japan 6.8% USA 6.2% UK 6.1% Germany 6.0% UAE 4.2%	
Exports:	Saudi Arabia 3.5% USA 2.5% UAE 2.5%	

Index of Globalisation: 60.93   **Rank:** 53

Military: **Active:**   8,200
                 Defence budget: $539 million FMA (USA) 15.7 million (0.4% of total)
                 % Middle East: 0.6

Index of Martial Potency: 4.04   **Rank:** 89

Political: **Type:** Constitutional Monarchy   **Independence:** 15 August 1971
                  Capital: Manama
                  Key individuals: al-Khalifa family

Freedom in the World Rating: **PR:** 5   **CL:** 5   **Status:** PF

Failed State Index: 57   **Rank:** 134

Recent events

1986   27-km King Fahd Causeway, linking Bahrain to Saudi Arabia, opened.
2001   Boundary settlement with Qatar.
2002   First parliamentary elections since 1973; boycotted by opposition.
2006   Parliamentary elections result in major gains for main Shi'a opposition party (Al Wefaq).

Key issues

Strategic location with regard to petroleum.
Increasing dependence upon Saudi Arabia.
Sunni–Shi'a balance among the Muslim population.
Trafficking of people.

Status

Located centrally in the Gulf between the major oil-producing states of Saudi Arabia, Iran and the UAE, Bahrain is in a highly strategic position. In addition, its near neighbour Qatar is a key source of natural gas. In sharp contrast to these states, the petroleum resources of Bahrain, which were exploited early, are in sharp decline and the economy has been increasingly built upon diversification and the provision of services. With an area of only 665 km² and a population of under three-quarters of a million, one-third of which is non-national, Bahrain is in a very real sense a micro-state. Of all the Middle Eastern states, only Mauritania has a lower GDP and defence budget. Thus Bahrain provides an obvious basis for comparison between the states of the region. However, despite the disadvantages of its size, Bahrain has an average per capita income which is larger than all the Middle Eastern states except Qatar, Kuwait and the UAE. It is a place of contrasts with more unrest than the other GCC states but also probably a higher level of democracy. Much of the unrest results from the sectarian balance in which the ruling family is Sunni but the Muslim balance of the country is 40 per cent Sunni to 60 per cent Shi'a. The success of the major Shi'a opposition party, Al Wefaq, in the 2006 election has yet to translate into concrete gains for Bahrain's disaffected Shi'a majority, and further sectarian unrest cannot be discounted.

On the broader political front, its location means that Bahrain is always likely to become implicated in any Gulf conflict over oil. Key international developments have been the settlement of the long-running boundary dispute with Qatar in 2001. The ICJ allocated sovereignty over Zubarah, Janan Island and the low tide elevation of Fasht ad Dibal to Qatar. Bahrain was granted sovereignty over the Hawar Islands and the Island of Qit'at Jaradah. The other key issue which represented geopolitical change was the opening of the 27-km long causeway, the King Fahd Causeway, with Saudi Arabia in 1986. Bahrain is already dependent upon the Kingdom for oil and water and the Causeway represents further dependence. For Saudi Arabia, the Causeway represents a potential problem in that Saudi citizens can sample the somewhat looser interpretation of Islam as practised in Bahrain.

Apart from more sophisticated transboundary movements, Bahrain is a destination for people trafficking from south and south-east Asia. These together with the women trafficked from Eastern Europe, Central Asia, Morocco and Thailand are liable to be subjected to conditions of involuntary servitude. Bahrain has made no progress in preventing trafficking and is listed as Tier Three.

CYPRUS

Location:	A relatively large island at the east end of the Mediterranean. Divided by a de facto boundary, the Attila Line, the area to the north being known, but not internationally recognised, as the Turkish Republic of Northern Cyprus (TRNC). (Map 33)

Area: 9,250 km² (3,355 km² TRNC)   **% Middle East:** 0.06
Comparison with Bahrain: 13.9×

Boundaries: **Land: length:** 150 km
Contiguous countries: UK Sovereign bases
Maritime: **Coastline:** 648 km
Claims: territorial sea 12 nml
contiguous zone 24 nml
continental shelf to 200 nml or depth of exploitation
Settlements: UK (Sovereign base areas) 1960 Sovereign bases: Akrotiri and Dhekelia

Boundary Vulnerability Index: 0.5
Geography: **Topography:** North mountain chain to terminus of the Karpas Peninsula, upland to south and west and central lowland.
Climate: Warm temperate Mediterranean.

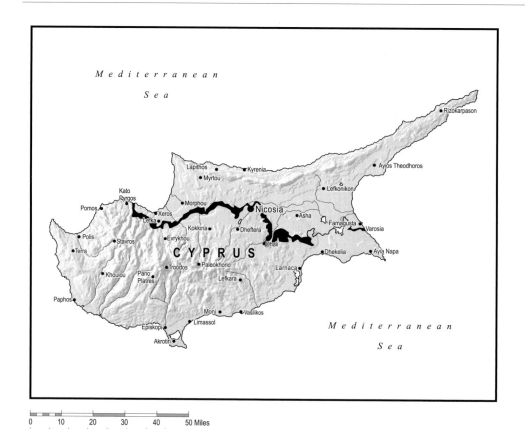

Map 33 Cyprus

Annual water withdrawal (domestic/industrial/agricultural): 0.21 cu km/yr
        (27%/1%/71%)   **per capita:** 250 cu m/yr (2000)

Population: 788,500
              % Middle East: 0.2   **Comparison with Bahrain:** 1.1×
              Literacy: 97.6%
              Ethnic groups: Greek 77% Turkish 18%

Economy: **GDP:** $25.95 billion   **% Middle East:** 0.6 (includes TRNC)
              Comparison with Bahrain: 1.05× **GDP per capita:** $27,100 (non-TRNC)
                                                    TRNC: $7,135

              Key resources: agricultural
              Cropped land: 15.13%   **Irrigated land:** 400 km²

Human Development Index Rank: 28

Culture: **Languages:**   Greek, Turkish, English
                Religion: Greek Orthodox 78% Muslim 18%
                Legal system: Based on English common law

Connectivity: **Telephones (fixed):** 494,500   **Mobiles:** 920,700
                Internet users: 356,600   **% of population:** 45
                Imports: Greece 17.6% Italy 11.4% Germany 9% UK 8.9% Israel 6.3%
                     France 4.3% Netherlands 4.3% China 4.2%
                Exports: UK 15.1% Greece 14.2% France 7.7% Germany 4.9% UAE 4.2%

Index of Globalisation: 62.48   **Rank:** 48

Military: **Active:**   10,000
                Defence budget: $346 million
                % Middle East: 0.4

Index of Martial Potency: 3.75   **Rank:** 98

Political: **Type:** Republic   **Independence:** 16 August 1960
                                                               (TRNC 1983)

                Capital: Nicosia
                Key individuals: Archbishop Makarios

Freedom in the World Rating: **PR:** 1   **CL:** 1   **Status:** F

Failed State Index: 70.2   **Rank:** 113

Recent events

1974   Military intervention by Turkey after a Greek government attempt to seize control of
        the island; the island divided by a de facto boundary.
1975   The Turkish area of Cyprus proclaimed self-rule.
1983   The TRNC declared and recognised by only Turkey.
2004   Referendum on reunification; voted down by Greek Cypriots.
        May: Greek-controlled Cyprus admitted as full member of EU.

Key issues

The continuing division of the island by the Attila Line between Greek and Turkish Cyprus.
Relationship with the EU, Greece and Turkey.
UK military bases.
Transit point for drug smuggling and people trafficking.

Status

Cyprus is the third largest island in the Mediterranean and, with Malta, one of the only two
sovereign states among Mediterranean islands. It is located at the eastern end of the Mediterra-

nean, effectively between Europe and Asia and between European and Middle Eastern culture. Cyprus has the highest literacy rate among Middle Eastern states. Prior to 1960, there was a strong movement for union with Greece, pursued by the National Organisation for Cypriot Struggle (EOKA). The *de facto* division of the island with regard to area is Greek 59 per cent to Turkish 37 per cent. The Attila Line occupies 4 per cent of the surface area. This areal division can be contrasted with the distinction between the ethnic groups which is Greek 77 per cent and Turkish 18 per cent. Thus, one continuing source of discontent has been the disproportionate area occupied by the Turkish Cypriots, given their proportion of the population. The separation of the population is by the Attila Line known as the Green Line, which has been maintained by UNIFCYP (UN Force in Cyprus).

On 1 May 2004 the entire island entered the EU although the EU "Acquis", the body of Common Rights and Obligations, applies only to the Greek Cypriot area, which is deemed to be under direct government control, and is suspended in the areas administered by the Turkish Cypriots. A long-running source of discontent has been the continuing presence of the UK sovereign base areas, Akrotiri and Dhekelia, both in the Greek Cypriot area. The latter extends from the Green Line to the sea thereby effectively cutting off an area of Greek Cyprus. These bases continue to be used by the British military for operations in the Middle East and Cyprus is thereby implicated in the various conflicts.

Given its location and strong links with both the EU and the Middle East, it is unsurprising that Cyprus has become a distribution point for illicit drugs including heroin and hashish and probably cocaine. It is also particularly vulnerable to money laundering. Cyprus is a destination and transport point for large-scale people trafficking from Eastern and Central Europe, the Far East, the Indian subcontinent and the Philippines. Many of these people are forced into involuntary servitude including sexual exploitation. Cyprus is on the Tier Two Watch List.

The further development of Cyprus must be related to reunification and, despite the fact that the Greek Cypriots rejected the UN settlement plan in their referendum taken in April 2004, recent developments look more promising. The election of Demetris Christofias in the run-off presidential election of February 2008 and the replacement of Rauf Denktash with new President Mehmet Ali Talat mean that both parts of Cyprus are now led by pro-settlement political leaders. Settlement talks between the two sides duly commenced in March 2008 but have yet to yield concrete dividends. Both Greece and Turkey have strengthened links with the island, the latter by the regular delivery of water under Operation Medusa.

EGYPT

Location: At the east end of North Africa abutting on to Asia with coastlines on the Mediterranean and Red Sea. (Map 34)

Area: 1,001,450 km²   **% Middle East:** 6.1
                      Comparison with Bahrain: 1,505.9×

Boundaries: **Land: length:** 2,665 km
        Contiguous countries: Palestine (Gaza Strip) 11 km, Israel 266 km,
              Libya 1,115 km, Sudan 1,273 km
        Maritime: **Coastline:**   2,450 km
              Claims: territorial sea 12 nml EEZ 200 nml
                    contiguous zone 24 nml
                    continental shelf to depth of 200 m or exploitation

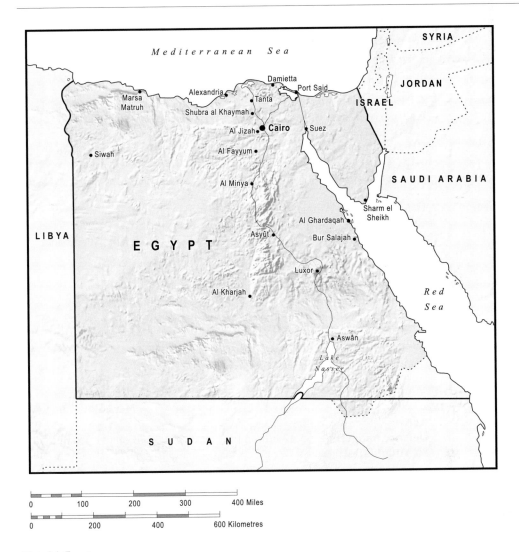

Map 34 Egypt

Settlements: Land: Gaza Strip: Camp David Accords 1979 ICJ Taba dispute 1988 Israel 1906, 1979, 1988 Libya 1925, 1926 Sudan 1899

Boundary Vulnerability Index: 7.5

Geography: **Topography:**   Hot desert plateau and plain with mountains and sand seas and the valley of the River Nile.

Climate: Hot and arid desert modified on coastal strip.

Annual water withdrawal (domestic/industrial/agricultural):   68.3 cu km/yr (8%/6%/86%)   **per capita:**   923 cu m/yr (2000)

Population: 80.3 million
% Middle East: 15.9   **Comparison with Bahrain:** 114.7×
Literacy: 71.4%
Ethnic groups: Egyptian 98%

Economy: **GDP:** $431.9 billion   **% Middle East:** 10.3
Comparison with Bahrain: 17.5×   **GDP per capita:** $5,400
Key resources: petroleum, natural gas, agricultural
Cropped land: 3.42%   **Irrigated land:** 34,220 km²

Human Development Index Rank: 112

Culture: **Languages:**   Arabic (official), English, French
Religion: Muslim (predominantly Sunni) 90%, Coptic 9%
Legal system: Based on Islamic and Civil law (especially Napoleonic codes)

Connectivity: **Telephones (fixed):** 10.808 million **Mobiles:** 18.001 million
Internet users: 6 million   **% of population:** 7.5
Imports: USA 11.4% China 8.3% Germany 6.6% Italy 5.4%
Saudi Arabia 5% France 4.6%
Exports: Italy 12.1% USA 11.3% Spain 8.7% UK 5.5% France 5.4%
Syria 5.1% Saudi Arabia 4.3% Germany 4.2%

Index of Globalisation: 54.18   **Rank:** 64

Military: **Active:**   468,500
Defence budget: $3.42 billion FMA (USA) 1.3 billion (31.4% of total)
% Middle East: 3.5

Index of Martial Potency: 6.19   **Rank:** 28

Political: **Type:** Republic   **Independence:** 28 February 1922
Capital: Cairo
Key individuals: Presidents Nasser, Sadat, Mubarak

Freedom in the World Rating: **PR:** 6   **CL:** 5   **Status:** NF

Failed State Index: 89.2   **Rank:** 36

Recent events

1956   Invasion by UK, France and Israel following the nationalisation of the Suez Canal.
1958   Joined with Syria to form the short-lived United Arab Republic (UAR).
1959   Camp David Accords signed with Israel and the USA; this brought a lasting end to conflict with Israel and resulted in Egypt's expulsion from the Arab League until 1989.
1981   Anwar Sadat assassinated.
1991   Gulf War; Egypt entered on the side of the Alliance with the third largest force.

Key issues

Ownership and operation of the Suez Canal, the most important such waterway in the world.
The isthmus of Suez is also the location of the Suez-Mediterranean oil pipeline (SUMED), a
    crucial oil pipeline, linking the Red Sea with the Mediterranean.
Hydropolitics of the Nile.
Unsettled land boundaries.
Centre for drug smuggling and people trafficking.
Continuing problems with terrorism and Islamic Fundamentalism.

Status

With its location, population, area and GDP, Egypt is a key Middle Eastern and the most important
Arab state. It has a population of over eighty million, almost ten million more than that of Turkey
and fifteen million more than that of Iran, the other Middle Eastern super-states. In area it is the
seventh largest state but, of those with a greater area, only Iran is comparable in population. Its
GDP accounts for 10 per cent of the regional GDP, marginally behind that of Saudi Arabia and
significantly less only than the GDPs of Iran and Turkey. Unlike Turkey and Iran, Egypt occupies
a central location in the Middle East. Through its ownership of the Sinai Peninsula, it controls the
land bridge between Africa and Asia. It is nodal between the *Mashreq* and the *Maghrib* and it
provides the link between the Middle East and Africa. In an essentially arid region, it is located in
the lower course of one of the world's great rivers and it controls a shipping artery of global significance, the Suez Canal. Furthermore, it abuts on to Israel, the one non-Arab and non-Islamic state
established controversially in the centre of the Arab and Islamic world.

It is therefore not surprising that Egypt has long been involved in pan-Arabism and a range of
movements inspired by Islam. Economically, Egypt has huge dependence upon the waters of the
River Nile and this has placed it centrally in the hydropolitics of the Nile basin. Disagreements
with Sudan have been settled but those with Ethiopia, which controls the headwaters of the Blue
Nile, the major source of water for the system, continue. The economy depends heavily upon
returns from the Suez Canal operation, tourism and remittances. There were severe economic
problems when the Canal was blocked during the 1956 and 1967 wars with Israel. Subsequent
enlargement of the canal and a change in the pattern of tanker use have restored the importance
of Suez. Tourism has undoubtedly been affected by bouts of terrorism. As the state with the
largest Arab population, a population which is relatively highly educated, Egypt has exported
expertise throughout the Arab world and benefited from the remittances. These have been
affected adversely by the various conflicts, most notably the Gulf War. However, following the
signing of the Camp David Accords (1978), Egypt has been in receipt of more US aid than any
other state except Israel.

Progress towards democratisation of the political system have been glacial, in part because
the most likely beneficiaries of truly free and fair elections would be the Muslim Brotherhood.
Though touted as a move in the direction of political reform, a series of amendments to the
constitution that were approved by referendum in 2007 are considered by most observers to
weaken civil and political rights and strengthen the executive at the expense of other branches
of government.

As a result of its highly strategic location, Egypt has become a key point for the transit of
drugs, particularly heroin, from south-west and south-east Asia to Europe and the USA. It is
heavily implicated in people trafficking and is on the Tier Two Watch List.

GAZA STRIP AND THE WEST BANK (PALESTINE)

Gaza Strip

Location: A coastal strip on the eastern Mediterranean at the southern end of the Levant coastline. Together with a small area around Jericho, it was transferred to the Palestinian Authority. (Map 35)

Area: 360 km²   **% Middle East:** 0.002
Comparison with Bahrain: 54%

Boundaries: **Land: length:** 62 km
Contiguous countries: Egypt 11 km, Israel 51 km
Maritime: **Coastline:** 40 km
Settlements: Status subject to Israeli–Palestine Interim Agreement

Boundary Vulnerability Index: 18

Geography: **Topography:**   Sandy coastal plain.
Climate: Hot and arid modified by coastal location.

Annual water withdrawal (domestic/industrial/agricultural): –
per capita: –

Population: 1.5 million
% Middle East: 0.3   **Comparison with Bahrain:** 2×
Literacy: 92.4%
Ethnic groups: Palestinian Arab

Economy: **GDP:** $5.034 billion*   **% Middle East:** 0.1
Comparison with Bahrain: 20%   **GDP per capita:** $1,100
Key resources: agricultural
Cropped land: 50%   **Irrigated land:** 150 km² *

Human Development Index Rank: 106 (for Occupied Territories)

Culture: **Languages:**   Arabic, Hebrew, English
Religion: Muslim (predominantly Sunni) 99.3%
Legal system: –

Connectivity: **Telephones (fixed):** 349,000*   **Mobiles:** 1.095 million*
Internet users: 243,000*   **% of population:** 6.0*
Imports: Israel, Egypt, West Bank
Exports: Israel, Egypt, West Bank

Index of Globalisation: –   **Rank:** –

Military: **Active:**   –
Defence budget: –
% Middle East: –

Map 35 Gaza Strip

Index of Martial Potency: –   **Rank:** –

Political: **Type:** –   **Independence:** –
                     Capital: Gaza City
                     Key individuals: Yasser Arafat

Freedom in the World Rating: **PR:** –   **CL:** –   **Status:** –

Failed State Index: –   **Rank:** –

Note: * includes West Bank.

West Bank

Location: Landlocked territory, to be negotiated between Palestine and Israel, lying
                     between Israel and Jordan on the west bank of the River Jordan. (Map 36)

Area: 5,860 km²   **% Middle East:** 0.04
                     Comparison with Bahrain: 8.8×

Boundaries: **Land: length:** 404 km
                     Contiguous countries: Jordan 97 km, Israel 307 km
                     Maritime: **Coastline:** Landlocked
                     Settlements: Status subject to Israeli–Palestine Interim Agreement

Boundary Vulnerability Index: 18

Geography: **Topography:**   Rolling plateau between the coastal plain of Israel and the
                                          Jordan Valley.
                     Climate: Mediterranean, modified by desert influences to the south.

Annual water withdrawal (domestic/industrial/agricultural): –
                                          per capita: –

Population: 2.5 million (in addition 187,000 Israeli settlers in West Bank and
                     approximately 177,000 in East Jerusalem)
                     % Middle East: 0.5   **Comparison with Bahrain:** 3.6×
                     Literacy: 92.4%
                     Ethnic groups: Palestinian Arab and other 83% Jewish 17%

Economy: **GDP:** $5.034 billion*   **% Middle East:** 0.1
                     Comparison with Bahrain: 0.2×   **GDP per capita:** $1,100*
                     Key resources: agricultural
                     Cropped land: 35.87%   **Irrigated land:** 150 km²*

Human Development Index Rank: –

Culture: **Languages:**   Arabic, Hebrew, English
                     Religion: Muslim (predominantly Sunni) 75% Jewish 17% Christian
                                          and other 8%

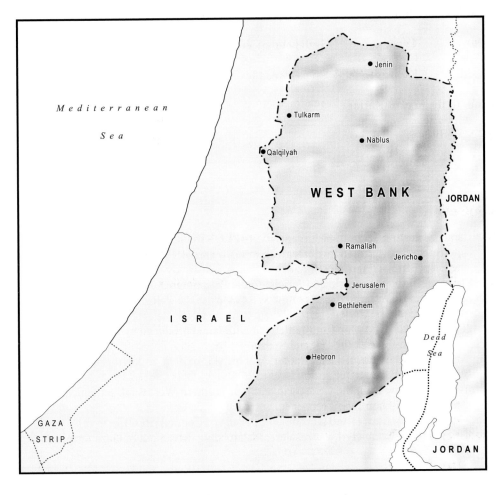

Map 36 West Bank

Legal system: –

Connectivity: **Telephones (fixed):** 349,000*   **Mobiles:** 1.095 million*
                     Internet users: 243,000*   **% of population:** 6.0*
                     Imports: Israel, Jordan, Gaza Strip
                     Exports: Israel, Jordan, Gaza Strip

Index of Globalisation: –   **Rank:** –

Military: **Active:**   –
                 Defence budget: –
                 % Middle East: –

Index of Martial Potency: –   **Rank:** –

Political: **Type:** –   **Independence:** –
                       Capital: –
                       Key individuals: Yasser Arafat

Freedom in the World Rating: **PR:** –   **CL:** –   **Status:** –

Failed State Index: –   **Rank:** –

Note: * includes Gaza Strip.

Recent events

1993   Israel–Palestine Liberation Organisation (PLO) Declaration of Principles (DOP) on interim self-government arrangements transitional to Palestinian self-government in the Gaza Strip and the West Bank.
1994   Transfer of the Gaza Strip and Jericho to the Palestinian Authority.
1995–9 Transfer of other limited areas of the West Bank to the Palestinian Authority.
2000   Second *Intifada.*
2003   The so-called Road Map for a settlement with Israel produced by the Quartet comprising the USA, the EU, the UN and Russia.
2005   Mahmud Abbas elected Palestinian Authority President following the death of the Palestinian leader Yasser Arafat in late 2004.
        Israel withdrew from the Gaza Strip and four northern West Bank settlements.
        Rafah crossing into Egypt reopened.
2006   Hamas won control of the Palestinian Legislative Council (PLC).
2007   Brief Hamas–Fatah civil war results in Fatah being driven out of Gaza Strip.

Key issues (Gaza Strip)

Israeli control of most access to the Gaza Strip including maritime entry and air space.
Israeli superior military presence and continuing military activities in the Gaza Strip.
Refusal of the international community to accept the election of Hamas because it does not recognise the state of Israel and will not renounce violence; this has resulted in financial embargos of the Palestinian Authority since Hamas took office.
According to UNRWA (2006) there were 994,000 refugees in the Gaza Strip.

Key issues (West Bank)

The construction of the Israeli Separation Barrier around the West Bank; location not consistent with accepted boundaries (the Green Line).
Since 1948, about 350 Peacekeepers from the UN Truce Supervision Organisation (UNTSO) monitor ceasefires, armistice agreements and incidents in the region.
Another issue is the number of Israelis: 280,000 in the West Bank and 180,000 in east Jerusalem.
Refugees estimated by UNRWA (2006) at 705,000.

Status

The Gaza Strip is a narrow, densely populated, arid coastal plain with an extreme shortage of water. It is approximately half the size of Bahrain but has twice the population. Legally, its status is subject to the Israeli–Palestine Interim Agreement. The West Bank is a significantly larger landlocked upland area on the west bank of the River Jordan. It has an area 8.8 times that of Bahrain and a population 3.6 times larger. Its status varies. Jericho and its immediate surroundings with the Gaza Strip was transferred to the Palestinian Authority (PA) in accordance with the Israel–PLO 4 May 1994 Cairo Agreement. Additional areas of the West Bank were also transferred according to the following Israel–PLO Agreements: 28 September 1995 Interim Agreement; 15 January 1997 Protocol concerning Re-Deployment in Hebron; 23 October 1998 Wye River Memorandum; and 4 September 1999 Sharm el-Sheikh Agreement. The West Bank is a rolling plateau between the coastal plain of Israel and the Jordan Valley. Like the Gaza Strip, its climate is dry Mediterranean but in the case of the West Bank the modification is by desert influences rather than the sea.

The literacy rate of the Gaza Strip and the West Bank is 92.4 per cent, a figure higher than that of any other Middle Eastern country except Cyprus, Israel and Kuwait. In contrast, the GDP of the two areas is approximately one-fifth of that of Bahrain and is the lowest in the Middle East. In the region, the per capita income exceeds only that of Afghanistan. The disparity between the educational level of the population and the GDP per capita must be related to the dominant role played by Israel in all the affairs of the prospective Palestinian state together with the virtually continuous conflict the two areas have endured. The major distinction between the two areas is illustrated by the ethnicity and religious adherence. The Gaza Strip is almost 100 per cent Arab, predominantly Palestinian, while in the West Bank Palestinian and other Arabs make up 83 per cent of the population, the remaining 17 per cent comprising Jewish settlers. Consequently the religion of the inhabitants of the Gaza Strip is almost wholly Muslim whereas in the West Bank it is 75 per cent Muslim, 17 per cent Jewish and 8 per cent Christian and other religions.

In June 2007, a brief civil war between Hamas and Fatah culminated in a decisive victory for Hamas in the Gaza Strip. Consequently, political control over Palestinian territory is now divided between Hamas (in the Gaza Strip) and Fatah (in the West Bank). This *de facto* division will greatly complicate efforts to negotiate a settlement with Israel because Israel refuses to deal with Hamas. Indeed, on 27 December 2008 Israel launched a full-blooded twenty-two-day attack on Gaza in an unsuccessful attempt to remove Hamas.

IRAN

Location:	At the eastern end of the Middle East, between Iraq and Afghanistan, with coastlines on the Gulf, the Gulf of Oman and the Caspian Sea. (Map 37)

Area: 1.648 million km² **% Middle East:** 10.1
Comparison with Bahrain: 2478.2×

Boundaries: **Land: length:** 5,440 km
Contiguous countries: Armenia 35 km, Turkey 499 km, Azerbaijan (including Nakhichevan) 611 km, Pakistan 909 km, Afghanistan 936 km, Turkmenistan 992 km, Iraq 1,458 km
Maritime: **Coastline:** 2,440 km (also Caspian Sea 740 km)

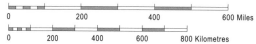

Map 37 Iran

Claims: territorial sea 12 nml
contiguous zone 24 nml
EEZ bilateral agreements or median lines in the Gulf
continental shelf to natural prolongation

Settlements: Bahrain 1971 Oman 1947 Qatar 1969 Saudi Arabia 1968 United Arab Emirates (Dubai) 1974

Boundary Vulnerability Index: 17.5

Geography: **Topography:** A mountainous rim with central desert basins and narrow coastal plains.

Climate: Hot desert modified by altitude and maritime influences.

Annual water withdrawal (domestic/industrial/agricultural): 72.88 cu km/yr (7%/2%/91%)   **per capita:** 1,048 cu m/yr

Population: 65.4 million

% Middle East: 13.0   **Comparison with Bahrain:** 93.4×
Literacy: 77%
Ethnic groups: Persian 51% Azeri 24% Gilaki and Mazandarani 8% Kurds 7% Arab 3%

Economy: **GDP:** $852.6 billion   **% Middle East:** 20.4
Comparison with Bahrain: 34.6× **GDP per capita:** $12,300
Key resources: petroleum, natural gas, coal, chromium and various metals
Cropped land: 11.07%   **Irrigated land:** 76,500 km²

Human Development Index Rank: 94

Culture: **Languages:** Persian and dialects 58% Turkic and dialects 26% Kurdish 9%
Religion: Muslim 98% (Shi'a 89% Sunni 9%)
Legal system: Based on Shari'a law

Connectivity: **Telephones (fixed):** 21.981 million   **Mobiles:** 13.7 million
Internet users: 18 million   **% of population:** 27.5
Imports: Germany 12.2% China 10.5% UAE 9.3% France 5.6% South Korea 5.4% Russia 4.4% Italy 5.4%
Exports: Japan 14.0% China 12.8% Turkey 7.2% Italy 6.3% South Korea 6.0% Netherlands 4.6%

Index of Globalisation: 35.19   **Rank:** 115

Military: **Active:**   545,000
Defence budget: $6.6 billion
% Middle East: 8.8

Index of Martial Potency: 6.72   **Rank:** 18

Political: **Type:** Theocratic Republic   **Independence:** Ancient State: Islamic Republic proclaimed 1 April 1979

Capital: Tehran
Key individuals: Ayatollah Khomeini, President Ahmadinejad

Freedom in the World Rating: **PR:** 6   **CL:** 6   **Status:** NF

Failed State Index: 82.8   **Rank:** 57

Recent events

1951 Nationalisation of the oil industry strongly opposed by the USA and the UK.

1975 Agreement reached with Iraq on a *thalweg* line for the Shatt al Arab boundary between the two states.

1979 Iranian revolution and the establishment of an Islamic theocratic republic under the guidance of Ayatollah Khomeini.

1979–81 US hostage crisis.

1980–8 War with Iraq which was supported by the West.

1986 Iran–Contra Scandal involving the administration of President Ronald Regan.

1987 Tanker war.

1988 Iran airbus A300B shot down by USA with loss of all 290 passengers and crew.

1989 Death of Ayatollah Khomeini.

1995 US embargo against Iran drafted.

2003 Commencement of sustained international interest in Iranian nuclear developments. US and UK military invasion of Iraq.

2005 President Ahmadinejad elected.

2009 President Ahmadinejad re-elected in disputed election. Large-scale popular protests result.

Key issues

Possible development of Iran as a nuclear power.

Dispute with the UAE over the ownership of the Tunbs Islands and Abu Musa. Potential dispute with Iraq over the Shatt al Arab boundary and its extension into the Gulf and with the littoral states over the delimitation of international boundaries in the Caspian Sea.

Threat of restricting or closing the Strait of Hormuz.

Hydropolitics of the River Tigris and flow to the Helmand River from Afghanistan.

Illegal transboundary movements, particularly the trafficking of people and the smuggling of drugs.

The siting of petroleum pipelines from the Caspian Sea Basin.

The global spread of Fundamentalist Islam and the potential incitement of terrorism.

Status

Iran occupies a highly strategic position, immediately beyond the Arab world, the eastern boundary of which is defined by its boundary with Iraq. It provides Middle Eastern links with the Trans-Caucasus, Central Asia and the Indian subcontinent. It also shares a boundary with the other major Middle Eastern non-Arab state, Turkey. Having land boundaries with seven states and actual or potential maritime boundaries with six further states, Iran has more contiguous neighbours than any other Middle Eastern state. As a result, there is a greater potential for conflict. With a coastline which stretches the entire length of the Gulf and the Gulf of Oman beyond, Iran has greater maritime access than any of the Gulf States and is able to command the Strait of Hormuz. Having a coastline on the Caspian Sea brings Iran into contact with Central Asia, the Trans-Caucasus and the Russian Federation. It is the state which links the oilfields of the Middle East with those of the Caspian Basin. Azeri and Kurdish minorities in the north-west of Iran involve the country in the problems of Azerbaijan and Kurdistan, the communities of Kurds found predominantly in Turkey and Iraq.

The boundary between Iran and Iraq represents the eastward limit of the Arab world and throughout history the Gulf has been the scene of conflict between Persians and Arabs. Indeed, there is still disagreement as to its correct name. It is known variously as the Persian Gulf, the Arab Gulf, the Persian–Arabian Gulf or the Gulf. The focus has been upon the boundaries, land and maritime, with Iraq and in particular upon the Shatt al Arab. Depending upon relative bargaining power, the boundary has changed from the *thalweg* of the Shatt al Arab to the eastern or Iranian bank. During the Gulf War, when Iraq was looking for support, agreement was reached on a *thalweg* boundary.

Most of Iran is arid but parts of the north and north-west receive considerable rainfall, up to 2,000 mm along stretches of the Caspian coast. Some 5 per cent of the country is irrigated and agricultural exports exceed imports. As the source of several major tributaries to the River Tigris, and the Karun which joins the Shatt al Arab, Iran is involved in the complex hydropolitics of the Tigris–Euphrates Basin. The foundation of the economy is however petroleum and it is this which allowed Iran the freedom of political action which it has enjoyed. Iran continues to act as a bastion against the encroachment of Western influences and derives increasing support from Russia and China. However, its ability as an actor on the world stage remains handicapped by the lack of any consistent, powerful, external support. Since it is feared by the Arab oil producers, Iran has had to obtain assistance of various kinds from an array of partners. However, it remains one of the three major powers of the Middle East and, after Saudi Arabia, a key global oil producer. Furthermore, natural gas resources are second only in the world to those of Russia. Given its location, Iran has been deeply involved in the Central Asian pipeline conundrum. Despite a degree of international isolation and the problems it has faced, Iran still boasts the second largest GDP in the Middle East, exceeded only marginally by the GDP of Turkey.

Notwithstanding the political turbulence, Iran is a largely literate, developed and relatively sophisticated society. Its ethnic makeup reveals the fact that only half the population is Persian; nearly a quarter is Azeri, while 7 per cent is Kurdish. These two minorities, concentrated in the north-west of the country, present recurrent problems.

According to the Index of Martial Potency, Iran is the fourth most powerful state militarily in the Middle East, marginally behind Turkey and Saudi Arabia but some distance behind Israel. It also has the fourth highest percentage of the Middle East defence budget, behind the same three states.

Politically, the modern state has been dominated by the teachings of Ayatollah Khomeini. The independent, hard line in foreign policy has continued with the election of the first non-cleric to the presidency, Mahmoud Ahmadinejad in 2005. From 1979, with rare intermissions, Iran has had to endure a degree of international isolation particularly following the emplacement of the US embargo from 1995. As a result, Iran has had to obtain arms predominantly from China and states of the former Soviet Union. A point of particular contention is the Iranian nuclear programme, discussed in more detail in Section E. Iran claims that the aim of the programme is for peaceful purposes but the USA and some other countries consider the ultimate goal to be the production of nuclear weapons. Iran has also been involved in numerous transboundary issues. It is heavily concerned in illegal drug trafficking and is a major trans-shipment node for the movement of heroin from south-west Asia to Europe. It is also a source, transit and destination country for the trafficking of women and girls for sexual exploitation and involuntary servitude in Pakistan, Turkey, the Gulf States and Europe and is listed as Tier Three.

Perhaps the major influence in the region results from Iran's mode of operation as a theocratic republic. In returning to what can be taken to be the basic fundamentals of Islam, the government has incited a good deal of violence and projected its concepts well beyond the boundaries of Iran. This has contributed significantly to the Islamic resurgence. It has also been related to the export of terrorism with strong fundamentalist implications, an example being indicated by the close links with Hezbollah. Ironically, the war in Iraq, pursued by the USA and the UK, has strengthened the political position of Iran because the Iraqi government is now dominated by Iran's Shi'a allies. Notably, the Islamic Supreme Council of Iraq (ISCI), the single most powerful political party in Iraq, was created and bankrolled by Iran in the 1980s during the Iran-Iraq War. Iran has also been accused of supporting the Sadrist movement, a mass movement that represents perhaps hundreds of thousands of Shi'a dispossessed in Iraq. The war in Afghanistan, which pits coalition forces against another Iranian enemy, the Taliban, seems likely to enhance Iran's position in the region still further.

The stability of the Islamic regime was tested in the immediate aftermath of the disputed presidential election in June 2009. Tens of thousands took to the streets of Tehran to protest against the re-election of President Ahmadinejad, claiming the result to be fraudulent. At the time of writing, the regime appears to have successfully re-established control and extinguished dissent.

IRAQ

Location:	At the head of the Gulf and comprises the middle and lower courses of the River Tigris and the River Euphrates. Virtually landlocked. (Map 38)

Area: 437,072 km²   **% Middle East:** 2.7
Comparison with Bahrain: 657.3×

Boundaries: **Land:** length: 3,650 km
Contiguous countries: Jordan 181 km, Kuwait 240 km, Turkey 352 km, Syria 605 km, Saudi Arabia 814 km, Iran 1,458 km

Maritime: **Coastline:** 58 km
Claims: territorial sea 12 nml
continental shelf not specified.

Settlements: Land: Iran 1847, 1913, 1937, 1975 Jordan informal Kuwait 1913, 1994 Saudi Arabia 1922, Syria 1916, 1920, 1932 Turkey 1925, 1926
Maritime: Kuwait 1993

Boundary Vulnerability Index: 12

Geography: **Topography:**   Low flood plain between the Zagros Mountains in the far north and the desert in the south.
Climate: Hot, arid desert, modified by altitude.

Annual water withdrawal (domestic/industrial/agricultural): 42.7 cu km/yr
(3%/5%/92%)   **per capita:** 1,482 cu m/yr (2000)

Population: 27.5 million
% Middle East: 5.5   **Comparison with Bahrain:** 39.3×
Literacy: 74.1%
Ethnic groups: Arab: 75-80% Kurdish: 15–20% Other (including Turkomans and Assyrians): 5%

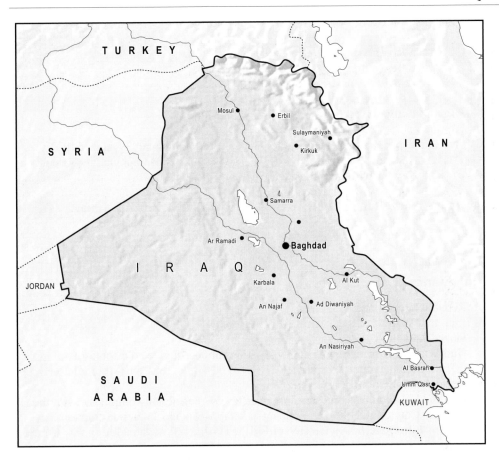

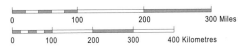

Map 38 Iraq

Economy: GDP: $100.0 billion   **% Middle East:** 2.4
Comparison with Bahrain: 4.9×   **GDP per capita:** $3,600
Key resources: petroleum, natural gas
Cropped land: 13.73%   **Irrigated land:** 35,250 km²

Human Development Index Rank: 135

Culture: **Languages:**   Arabic, Kurdish, Assyrian, Armenian
Religion: Muslim 97% (Shi'a 60–65% Sunni 32–37%)
Legal system: Based upon European civil and Islamic law

Connectivity: **Telephones (fixed):** 1.547 million **Mobiles:** 10.9 million
Internet users: 36,000   **% of population:** 0.1
Imports: Syria 26.5% Turkey 20.5% USA 12% Jordan 7.2%
Exports: USA 46.7% Italy 10.7% Canada 6.2% Spain 6.2%

Index of Globalisation: –   **Rank:** –

Military: **Active:**   494,800
                      Defence budget: –
                      % Middle East: –

Index of Martial Potency: 6.16   **Rank:** 31

Political: **Type:** Parliamentary democracy   **Independence:** 3 October 1932
                      Capital: Baghdad   (28 June 2004 Iraqi Interim Government)
                      Key individuals: President Saddam Hussein

Freedom in the World Rating: **PR:** 6   **CL:** 5   **Status:** NF

Failed State Index: 111.4   **Rank:** 2

Recent events

1958   Republic proclaimed.
1980–8   Iran–Iraq War.
1990   Seized Kuwait.
1991   Gulf War and Kuwait liberated; UN sanctions imposed; given the animosity of most of
          the surrounding countries and the very short coastline, sanctions were relatively easy
          to impose and monitor.
2003   US- and UK-led invasion ostensibly over Iraq's alleged non-compliance with the UN
          Security Council (UNSC) requirement to scrap weapons of mass destruction.
2004   Iraqi Interim Government took over from the Coalition Provisional Authority to govern
          under the Transitional Administrative Law (TAL) for Iraq.
2005   275-member Transitional National Assembly (TNA) was elected and the Iraqi Transi-
          tional Government (ITG) assumed office.
          Permanent Constitution approved.
          275-member Council of Representatives elected.
2006   20 May: constitutional government established.
2008   November: status of Forces Agreement concluded with the USA; requires the USA to
          withdraw all forces by 2011.
2009   Governorate elections result in big gains for Prime Minister Nuri al-Maliki's "State of
          Law" coalition.

Key issues

Coalition (US- and UK-led) forces in Iraq under UNSC mandate. Following extreme violence
     since 2003 the population of Baghdad has largely dispersed into separate Sunni and Shi'a
     enclaves; by 2008 the violence had reduced.
Status of the Kurds and Kurdistan: particular problem over the boundary of a recognised Kurdish
     area and whether or not Kirkuk should be included; continuing problems of Turkish inva-
     sions in northern part of the Kurdish area in pursuit of the Kurdish Workers' Party (PKK).
Hydropolitics of the Tigris–Euphrates: Iraq occupies the greater part of Mesopotamia between
     the two rivers but controls no headwaters of major rivers or tributaries.
Third River Project: the draining of the southern marshlands.

International boundaries: maritime boundaries with Kuwait and Iran to be negotiated; problems likely to remain over the Shatt al Arab boundary.

Refugees: 2 million Iraqis fled during and after the invasion, primarily to Syria and Jordan.

Iraq decimated by conflict and terrorism; problems include lack of basic services, fractured infrastructure and food security.

Status

Iraq, a former and future major oil producer, occupies a strategic location at the head of the Gulf. While it is essentially in the core of the Middle East, it is at the eastern edge of the Arab world. It abuts on to two of the three major Middle Eastern powers: Turkey and Iran. There are seven contiguous countries and over the past twenty years Iraq has had disputes, ranging from major wars to disagreements, with all of them. The one neighbour that has provided continuous support, although laced with criticism, is Jordan. A major geographical characteristic of its location is the fact that it is virtually landlocked with a coastline of only 58 km. This situation has undoubtedly exacerbated conflict with Iran and Kuwait while constraining Iraq's own political manoeuvrability.

Iraq is a medium-sized state in both area and population with a high Boundary Vulnerability Index. To the general political, social and economic chaos in the country since 1991 can be attributed its low literacy rate and very low GDP per capita. The latter at $3,600 is extraordinary for a major oil producer and exceeds that of only Sudan, Yemen, Mauritania, Palestine (Gaza Strip and West Bank) and Afghanistan among Middle Eastern countries. The lowest GDP per capita among other major oil producers is over 3.4 times as large. Heading the other problems is the fact that the Iraqi population is far from homogeneous. The Kurdish minority accounts for about 15–20 per cent of the population and there are several other significant minorities. Arabic is the most important language but there are several others spoken. Although Muslims make up 97 per cent of the population, there is a sharp split between the Shi'a who comprise 60–65 per cent and Sunni who account for 32–37 per cent.

Until the invasion, the government and state infrastructure had been run by Arab Sunnis. The Constitutional Government is dominated by an alliance of Kurdish parties and Shi'a religious parties closely affiliated with Iran. It has proven difficult to reconcile the Arab Sunni community to their new status as a political minority, and this is likely to create major problems for Iraq following the departure of the occupying forces. Other problems include the status of the oil-rich northern city of Kirkuk, and whether the management of the oil and gas sector is to be controlled by the federal government or the regions.

The future of Iraq is extremely difficult to forecast. All current signs indicate that it is unlikely to hold together as a single state and, at best, it seems possible it will become a deeply divided federal unit. The level of autonomy of the Kurdish region and the boundary of such a region are particularly contentious issues. Iraq is currently rated second on the Failed State Index.

ISRAEL

Location: East Mediterranean coastal plain, upland in the east and desert in the south. Coastlines on the Mediterranean and, minutely, the Red Sea. (Map 39)

Area: 20,770 km²   **% Middle East:** 0.1
Comparison with Bahrain: 31.2×

Map 39 Israel

Boundaries: **Land:** length: 1,017 km
  Contiguous countries: Syria 76 km, Lebanon 79 km, Jordan 238 km,
        Egypt 266 km, Palestine (Gaza and the whole of the
        West Bank) 358 km
  Maritime: **Coastline:** 273 km
    Claims: territorial sea 12 nml
        continental shelf to depth of exploitation
  Settlements: Land: Egypt 1906, 1979 (Camp David Accords) 1988
        (Taba) Gaza Strip 1979 Jordan 1994 Lebanon 1916, 1922,
        1939 Syria 1916, 1920, 1949
        Maritime: Jordan 1996

Boundary Vulnerability Index: 13.5

Geography: **Topography:**   Coastal plain and rolling upland inland to the Jordan/Dead
        Sea Valley with desert plain to the south.
  Climate: Mediterranean, modified by altitude, hot and arid in the
        south and east.

Annual water withdrawal (domestic/industrial/agricultural): 2.05 cu km/yr
  (31%/7%/62%)   **per capita:** 305 cu m/yr (2000)

Population: 6.4 million
  % Middle East: 1.3   **Comparison with Bahrain:** 9.3×
  Literacy: 97.1%
  Ethnic groups: Jewish 76.4% non-Jewish 23.6%

Economy: **GDP:** $184.9 billion   **% Middle East:** 4.4
  Comparison with Bahrain: 7.5×   **GDP per capita:** $28,800
  Key resources: agricultural
  Cropped land: 19.33%   **Irrigated land:** 1,940 km²

Human Development Index Rank: 23

Culture: **Languages:**   Hebrew (official), Arabic, English
  Religion: Jewish 76.4% Muslim 16% Arab Christian 1.7% Druze 1.6%
  Legal system: Mixture of English common law, British Mandate regula-
        tions and, in personal matters, Jewish, Christian and
        Muslim legal systems.

Connectivity: **Telephones (fixed):** 3.005 million   **Mobiles:**   8.404 million
  Internet users: 1.9 million   **% of population:** 29.7
  Imports: USA 12.4% Belgium 8.2% Germany 6.7% Switzerland 5.9%
        UK 5.1% China 5.1%
  Exports: USA 38.4% Belgium 6.5% Hong Kong (China) 5.9%

Index of Globalisation: 70.83   **Rank:** 29

Military: **Active:**   176,500
                Defence budget: $9.45 billion FMA (USA) 2.34 billion (56.5% of total)
                % Middle East: 9.7

Index of Martial Potency: 7.67   **Rank:** 8

Political: **Type:** Parliamentary democracy   **Independence:** 14 May 1948
                Capital: Tel Aviv (Jerusalem claimed by Israel)
                Key individuals: Prime Ministers Meir, Rabin and Sharon

Freedom in the World Rating: **PR:** 1   **CL:** 2   **Status:** F

Failed State Index: 79.6   **Rank:** 75

Recent events

Wars against Arab neighbours: 1948–9, 1956, 1967 and 1973; as a result of the 1967 war, Israel occupied the Golan Heights (Syria), the West Bank (Jordan), the old city of Jerusalem and the Gaza Strip (Egypt).

1967   UN Security Council Resolution 242 adopted. Called on Israel to withdraw "from territories occupied in the recent conflict".
1978   Camp David Accords signed. Followed, in 1979, by a Peace Treaty between Egypt and Israel.
1982   Invasion of Lebanon resulting in the long-term occupation of southern Lebanon. Withdrawal from the Sinai Peninsula.
1987   The *Intifada* focused global attention on the plight of the Palestinians and the continuing Israeli retention of the Occupied Territories; this brought worldwide sympathy for the Palestinian cause which was enhanced in 1988 by the decision of King Hussein to withdraw Jordan's claims to the West Bank.
1991   During the Gulf War, Iraq fired missiles at Israeli cities in an attempt to provoke retaliation.
1993   The Oslo Accords brought some agreement between Israel and the PLO with a declaration of principles.
1994   The Gaza–Jericho Agreement giving self-rule to those areas.
2000   Israel withdrew from southern Lebanon.
2003   The Quartet (the USA, the EU, the UN and Russia) produced the so-called Road Map.
2005   Israel officially disengaged from the Gaza Strip.
        Prime Minister Ariel Sharon formed new centrist Kadima Party. He was joined by prominent Labour Party figure Shimon Peres. Kadima became largest single party in Knesset after 2006 elections.
2006   The election of Hamas in the Gaza Strip froze relations between the Palestinians and Israel.
        War in Lebanon in which Israel was effectively defeated by Hezbollah and forced to withdraw.
2007   Annapolis Conference called by President George W. Bush with the optimistic aim of reaching a settlement between Israel and the Palestinians by the end of 2008.
2008   Abortive attempt to remove Hamas by military force.
2009   Parliamentary elections produce victory for right-wing political forces. Benjamin Netanyahu forms right-wing coalition government.

Key issues

The Occupied Territories: Israel has the following settlements and civilian land use sites: 242 in
the West Bank, 42 in the Golan Heights and 29 in east Jerusalem.

There are 280,000 Israeli settlers in the West Bank, 18,000 in the Golan Heights and 180,000 in
east Jerusalem.

Land and peace: a key question is whether Israel should relinquish the Occupied Territories in
return for guaranteed peace.

Land and water: Israel, the Gaza Strip, the West Bank and Jordan all have acute water shortage
problems. If Israel were to relinquish the West Bank, it would lose part of its access to the
Jordan River but, more important, use of the West Bank aquifers, a major contributor to
Israeli water supplies.

Hydropolitics of the Jordan Basin: Syria, Jordan, Israel and the West Bank are all intimately
involved in the distribution of water from the River Jordan. Only one of the three tributaries of the river, the Dan, rises in Israel.

Long-term security: despite strong and virtually uncritical support from the USA, Israel
comprises a population of only 6.4 million of which almost a quarter is non-Jewish and
mostly Arab. Israel is effectively opposed by the entire Arab world, although both Egypt
and Jordan have signed peace deals officially recognising Israel's existence.

Zionism: while Israel must consider giving up territory, there is pressure from Zionist fundamentalists to expand and occupy the entire Land of the Bible.

Nuclear weapons: Israel has a known nuclear weapons production facility at Dimona and is
thought to have of the order of 100–300 nuclear devices but it has not signed the Nuclear
Non-proliferation Treaty (NPT).

Area: with an area of only 0.1 per cent of the Middle East, Israel is little more than a micro-state
and could not afford a war of attrition.

Money laundering: drugs trafficking particularly from Lebanon and Jordan and people trafficking from Egypt.

Status

Israel occupies a strategically important location regionally in that it is in the centre of the Arab
world and, more locally, because it has access to the Mediterranean and the Red Sea and Suez
Canal. It is unique in the world, having been founded upon a specific religion, race and culture.
Apart from Cyprus, it is the only non-Islamic state in the Middle East. Following the Balfour
Declaration of 1917 and the partitioning of Palestine by a narrow vote in the UN General Assembly in 1947, Israel was implanted as the one geographical break in the otherwise continuous
Arab world. The area awarded formed a significant part of the Land of the Bible but, within the
original borders, barely one-third of the population was Jewish. At some stage all the surrounding states have lost territory to Israel but the major loss was sustained by the Palestinians, originally the dominant population of the area. Both the fact and the manner of its establishment have
resulted in almost continuous conflict since independence in 1948.

The state of Israel within its 1967 borders is almost a micro-state but with a relatively large
population. Its population exceeds that of Jordan, the area of which is some 4.5 times larger. It
has four contiguous countries together with the future state of Palestine and, since 1948 has been
at war at some time with all of its neighbours.

There are virtually no resources other than agricultural and for the production of these Israel
is handicapped by major water deficiencies. Nonetheless, aided by global and particularly US

financial support, Israel has developed an economy with a GDP which is among the highest in the Middle East, exceeded only by those of Iran, Turkey, Saudi Arabia, Egypt and Algeria, all mega-states and vastly superior in area, population and resources. Apart from the Qatar, Kuwait and the UAE, all petroleum-rich states with very small populations, Israel has the highest GDP per capita in the Middle East. Its literacy rate at 97.1 per cent is the second highest in the Middle East, exceeded fractionally by that of Cyprus. The Index of Martial Potency and the rank of eighth in the world illustrate the importance attached to security. Israel's defence budget is the fourth highest in the Middle East behind those of Saudi Arabia, Turkey and the UAE. It is almost three times that of Egypt and a quarter larger than that of Iran. Israel also receives 56.5 per cent of the US military assistance dispersed in the region. If this is added to the defence budget, the total is second only to the defence budget of Saudi Arabia.

Given the disparity in population size between Israel and its neighbours and the level of hostility towards it in the Arab world in general, the importance of Israel's support from the USA cannot be overstressed. Should that waver before an agreed peace with its neighbours has been established, the future of Israel would look very bleak. The possession of nuclear weapons would only postpone the inevitable and could result in the despoliation of large areas of the Middle East. For Israel, guaranteed peace must be the fundamental aim.

Though Israel has comfortably the highest level of human development in the region, the relative affluence of Israeli citizens stands in stark contrast to the desperate plight of those in the Occupied Territories. Among developed countries, Israel is second only to the USA in terms of income inequality. Partly as a result of the dire conditions in the West Bank, Israel moved into the top sixty countries on the Failed State Index for the first time in 2008.

JORDAN

Location:	To the east of Israel, almost landlocked, between the Jordan Valley and the Arabian Desert. (Map 40)

Area: 92,300 km² **% Middle East:** 0.6
Comparison with Bahrain: 138.8×

Boundaries: **Land: length:** 1,635 km
Contiguous countries: Palestine (West Bank) 97 km, Iraq 181 km, Israel 238 km, Syria 375 km, Saudi Arabia 744 km
Maritime: **Coastline:** 26 km
Claims: territorial sea 3 nml
Settlements: Land: Iraq informal, Israel 1994 Saudi Arabia 1925, 1965 Syria 1916, 1920, 1931, 2004
Maritime: Israel 1996

Boundary Vulnerability Index: 4

Geography: **Topography:** Uplands along the Jordan Valley with desert plateaux and plains to the east.
Climate: Hot and arid desert.

Annual water withdrawal (domestic/industrial/agricultural): 1.01 cu km/yr
(21%/4%/75%) **per capita:** 177 cu m/yr

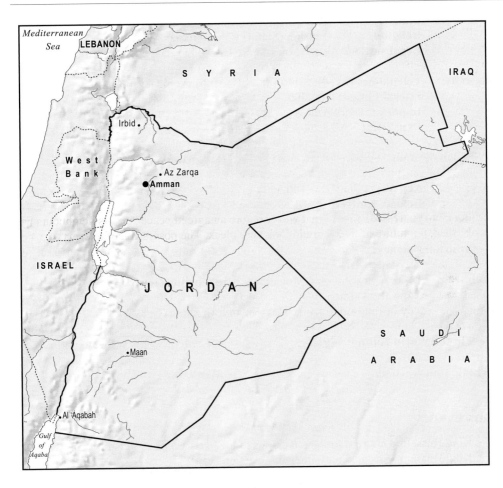

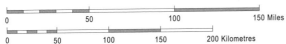

Map 40 Jordan

Population: 6.1 million
%**Middle East:** 1.2   **Comparison with Bahrain:** 8.6×
Literacy: 89.9%
Ethnic groups: Arab 98%

Economy: **GDP:** $28.2 billion   %**Middle East:** 0.7
Comparison with Bahrain: 1.1×   **GDP per capita:** $4,700
Key resources: agricultural, chemical
Cropped land: 4.5%   **Irrigated land:** 750 km²

Human Development Index Rank: 86

Culture: **Languages:**   Arabic (official), English
Religion: Muslim (Sunni) 92% Christian 6%
Legal system: Based on Islamic law and French codes

Connectivity: **Telephones (fixed):** 614,000   **Mobiles:** 4.343 million
Internet users: 707,000   **% of population:** 13.0
Imports: Saudi Arabia 23.2% Germany 8.3% China 8.0% USA 5.3%
Exports: USA 25.2% Iraq 16.9% India 8.0% Saudi Arabia 5.8% Syria 4.7%

Index of Globalisation: 64.74   **Rank:** 39

Military: **Active:**   100,500
Defence budget: $1.59 billion FMA (USA) 206 million (5.0% of total)
% Middle East: 1.6

Index of Martial Potency: 5.59   **Rank:** 39

Political: **Type:** Constitutional Monarchy   **Independence:** 25 May 1946
Capital: Amman
Key individuals: Kings Hussein and Abdullah II

Freedom in the World Rating: **PR:** 5   **CL:** 4   **Status:** PF

Failed State Index: 76.6   **Rank:** 82

Recent events

1988   King Hussein withdrew Jordanian claims to the West Bank.
1989   King Hussein reinstated parliamentary elections.
1991   During the Gulf War, Jordan's official neutrality and apparent support for Iraq alienated Saudi Arabia and Kuwait.
1994   Peace Treaty signed with Israel.
1999   King Hussein died and was succeeded by his son, King Abdullah II.

Key issues

Relations with Iraq and the Palestinians.
Boundary issues with Palestine (the West Bank) and Israel; 2004 boundary settlement with Syria.
Refugees from Palestine, Kuwait and Iraq.
Economic dependence upon remittances, transit trade and aid.

Status

Jordan is a small state located in the cockpit of the Middle East with five contiguous countries: Palestine (West Bank), Iraq, Israel, Syria and Saudi Arabia. It is therefore affected by most Middle Eastern turbulence. Like Iraq, it is virtually landlocked with only 26 km of coastline, however it does have access to the strategic Gulf of Aqaba. At times, its supportive relationship with Iraq has resulted in discord with other Arab countries such as Saudi Arabia and the discontinuation of aid. Jordan has been the most consistent supporter of the Palestinians but both King Hussein and now King Abdullah II have used skilful diplomacy in fostering generally good relationships on all sides. The main resource of Jordan is agricultural products although there are also chemicals but agriculture is severely limited by the shortage of water; most is focused along the Jordan Valley. The other resource is the population and the literacy rate of 89.9 per cent is higher than that of any other Middle Eastern state except for Cyprus, Israel, Kuwait and the future state of Palestine. With very limited resources, it has achieved a GDP per capita that is significantly higher than that of both Egypt and Syria.

Potential for instability in Jordan has increased recently owing to terrorist attacks on Jordanian soil, most notably, the multiple suicide bombings of hotels in Amman in November 2005, and the influx of up to a million Iraqi refugees escaping violence in their homeland.

The future development of Jordan depends upon the goodwill of other Arab states, particularly the major oil producers. It is particularly important that King Abdullah II manages to replicate the diplomatic balancing act of his father. That is, to placate a restive majority Palestinian population while continuing to maintain good relations with Israel and the USA.

KUWAIT

Location: A micro-state located on the western side at the head of the Gulf. (Map 41)

Area: 17,820 km² **% Middle East:** 0.1
Comparison with Bahrain: 26.8×

Boundaries: **Land: length:** 462 km
Contiguous countries: Saudi Arabia 222 km, Iraq 240 km
Maritime: **Coastline:** 499 km
Claims: territorial sea 12 nml
Settlements: Iraq 1993 Saudi Arabia 2000

Boundary Vulnerability Index: 7.5

Geography: **Topography:** Desert plain with low ridge in the north.
Climate: Arid, hot desert with slight seasonal variation.

Annual water withdrawal (domestic/industrial/agricultural): 0.44 cu km/yr
(45%/2%/52%) **per capita:** 164cu m/yr (2000)

Population: 2.5 million (estimated 1.3 million non-nationals)
% Middle East: 0.5 **Comparison with Bahrain:** 3.6×
Literacy: 93.3%
Ethnic groups: Kuwaiti 45% Other Arab 35% South Asian 9% Iranian 4%

Economy: **GDP:** $138.6 billion **% Middle East:** 3.3
Comparison with Bahrain: 3.1× **GDP per capita:** $55,300
Key resources: petroleum
Cropped land: 1.01% **Irrigated land:** 130 km²

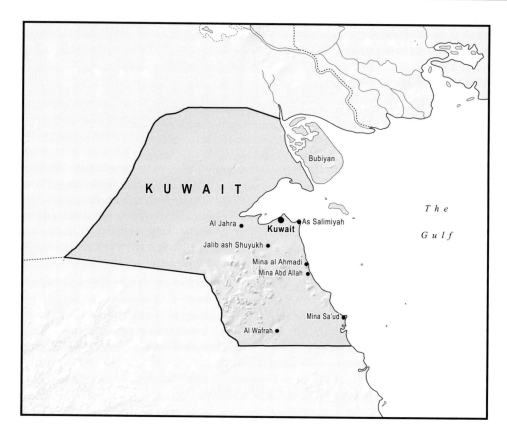

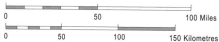

Map 41 Kuwait

Human Development Index Rank: 33

Culture:	**Languages:**	Arabic (official) (English commonly spoken)
	Religion:	Muslim 85% (Sunni 70% Shi'a 30%)
	Legal system:	Civil law with Islamic law in personal matters.

Connectivity: **Telephones (fixed):** 510,300   **Mobiles:** 2.536 million
Internet users: 817,000   **% of population:** 32.7
Imports: USA 14.1% Germany 7.9% Japan 7.8% Saudi Arabia 6.8%
China 5.7% UK 5.4% Italy 4.6%
Exports: Japan 20.4% South Korea 16.2% Taiwan 10.8% Singapore 9.7%
USA 9.0% Netherlands 5.3% China 4.1%

Index of Globalisation: 63.51   **Rank:** 43

Military:	Active:	15,500
	Defence budget:	$4.51 billion
	% Middle East:	6.0

Index of Martial Potency: 5.81   **Rank:** 37

Political:	**Type:** Constitutional Emirate	**Independence:** 19 June 1961
	Capital: Kuwait City	
	Key individuals: Al-Sabah family	

Freedom in the World Rating: **PR:** 4   **CL:** 5   **Status:** PF

Failed State Index: 62.1   **Rank:** 124

Recent events

1980–8   Iran–Iraq War in which Kuwait supported Iraq.
1990   2 August: attacked and occupied by Iraq. Liberated by US-led UN Coalition Force in a conflict which lasted 100 hours from 23 February 1991 (Operation Desert Storm). Severe damage to the oil infrastructure caused on departure of the Iraqi troops.
1993   Land boundary with Iraq redrawn under auspices of UN Commission. Maritime boundary delimited as far as the mouth of the Khor Abdulla.
         Large US military presence remains as a safeguard but implicates Kuwait in the current war on terror and specifically the war in Iraq.

Key issues

Located centrally between the major states of Iraq, Iran and Saudi Arabia with, until the current war, continuing intermittent threats from Iraq.
The Rumaila oilfield, a megafield, lies across the land boundary with Iraq.
Very limited natural water resources.
Major global oil producer: reserves approximate to those of Iran, Iraq and UAE and are second only to those of Saudi Arabia in the region.
In the predominantly Muslim (85 per cent population), the balance is Sunni (70 per cent) and Shi'a (30 per cent).

Status

Kuwait is a micro-state. Among the states of the Middle East, only Bahrain, Lebanon and Qatar are smaller in area, while only Qatar, Cyprus and Bahrain have a lower population. The largest state, Sudan, is 140 times greater in area and the most populous, Egypt has over thirty times the population of Kuwait. As a result, for security Kuwait is always likely to be dependent upon external assistance. Through its location, it cannot avoid becoming enmeshed in the many tensions and conflicts of the region. This fact can also be related to the size of its defence budget which, within the region, is exceeded only by those of Saudi Arabia, Turkey, the UAE, Israel and Iran.

The northern land boundary and the maritime boundary must be classed as unstable owing to long-term and continuing threats from Iraq, halted probably temporarily by the current war. Apart from local issues such as farming land, the location of the Rumaila oilfield, and therefore the potential for extraction from it, will remain a focus of contention on the land boundary. In the longer term, the maritime boundary may be affected by sea level change but the immediate prospect is likely to be continuing harassment from Iraq. Kuwait's major islands, Warbah and Bubiyan, limit maritime access to Iraq which, with only 58 km of coastline, is virtually land-locked. The maritime exit points for Iraq are through the ports of the Shatt al Arab on the eastern side and from Umm Qasr on the western side. Traffic to and from Umm Qasr needs to partially circumnavigate Warbah Island and this sea-lane vulnerability has resulted in a range of threats and requests from Iraq. These vary from the occupation of the country in 1990 to requests for long-term rental. Such issues arise periodically as, depending upon relations with Iran, the focus of Iraq's maritime activity shifts from east to west.

The economy of Kuwait is dominated by two natural resources. The rainfall (10–370 mm) is extremely low, there is no surface water and groundwater resources are very limited. Thus, there is a high dependence upon desalination for water supplies and this reliance upon a few large desalination plants represents an obvious security weakness. However, possibly because of the general vulnerability in the region, there are few records of attacks upon water sources. The possession of major oilfields is both a blessing and a burden. The oil has allowed a modern society to develop and flourish in what is otherwise an inhospitable environment but dependence upon a one-product economy brings its own problems. There have been efforts to diversify economically but the future will be secured more by the large-scale investments made possible by oil than by manufacturing. More important, as oil becomes scarcer globally, the pressure upon the major producers will grow. While this will improve the bargaining position of Kuwait, it must also result in enhanced military threats. Owing to the scarcity of water, a very low percentage of the land is cropped and an even smaller area irrigated. This results in a further vulnerability: the heavy dependence upon foreign sources for food supplies.

Oil has facilitated the development of an educated and relatively sophisticated society as indicated by a literacy rate of 93.3 per cent and a high level of Internet use. Kuwait has a very high per capita income. This last figure also contributes to Kuwait's relatively high Index of Globalisation. Services are generally of Western standards and Kuwait is among the more liberal of Gulf States. However, future tensions could arise through the Sunni–Shi'a balance. Should southern Iraq become essentially Shi'a, then given the preponderance of Shi'a in the eastern province of Saudi Arabia, there could be effectively a Shi'a curtain or crescent around the Gulf. The other key point about the population is that under half are Kuwaiti. There has long been a dependence upon other Arabs, particularly Palestinians, for the labour force. Following the events of 1990 and support shown for Saddam Hussein in Jordan, Palestinians were expelled from Kuwait and remain unwelcome. This has resulted in a variety of labour problems. A further significant point is the distribution of population which is located almost exclusively in the Kuwait City urban concentration and a number of other largish settlements along the coast to the south. To the north and west, most of the country remains virtually unpopulated. Kuwait is a destination and transit point for south, south-eastern and eastern Asian migrants and workers who are subject to involuntary servitude. Kuwait is on the Tier Two Watch List.

As a constitutional Emirate, political life in Kuwait is dominated by the Al-Sabah family. Kuwait appears more democratic than any of the other GCC countries, except perhaps Bahrain, in that there have been limited elections, but the ultimate power still resides with the extended family. The stability which has resulted is indicated by Kuwait's low position on the Failed State Index. However, the limited democracy has resulted in an ambivalent Freedom in the World Rating.

LEBANON

Location: East Mediterranean coast towards the northern end of the Levant coastline. (Map 42)

Area: 10,400 km²   **% Middle East:** 0.06
Comparison with Bahrain: 15.6×

Boundaries: **Land: length:** 454 km
Contiguous countries: Israel 79 km, Syria 375 km
Maritime: **Coastline:** 225 km
Claims: territorial sea 12 nml
Settlements: Land: Israel 1916, 1922, 1939 Syria 1920, 1945

Boundary Vulnerability Index: 10

Geography: **Topography:** Two parallel mountain ranges north-east to south-west with the Bekaa Valley between.
Climate: Mediterranean modified by altitude.

Annual water withdrawal (domestic/industrial/agricultural): 1.38 cu km/yr
(33%/1%/67%)   **per capita:** 385 cu m/yr (2000)

Population: 3.9 million
% Middle East: 0.8   **Comparison with Bahrain:** 5.6×
Literacy: 87.4%
Ethnic groups: Arab 95% Armenian 4%

Economy: **GDP:** $40.65 billion   **% Middle East:** 1.0
Comparison with Bahrain: 1.7×   **GDP per capita:** $10,400
Key resources: agricultural
Cropped land: 30.1%   **Irrigated land:** 1,040 km²

Human Development Index Rank: 88

Culture: **Languages:** Arabic (official), French, English, Armenian
Religion: Muslim 59.7% Christian 39%
Legal system: Mixture of Ottoman law, Canon law, Napoleonic code and civil law

Connectivity: **Telephones (fixed):** 681,000   **Mobiles:** 1.103 million
Internet users: 950,000   **% of population:** 24.4
Imports: Syria 11.6% Italy 9.8% USA 9.3% France 7.7% Germany 6.2% China 5.0% Saudi Arabia 4.7%
Exports: Syria 26.8% UAE 12.0% Switzerland 6.0% Saudi Arabia 5.7% Turkey 4.5%

Index of Globalisation: –   **Rank:** –

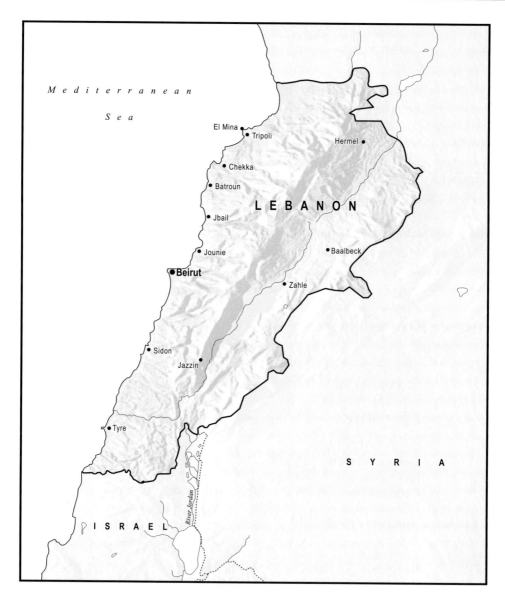

Map 42 Lebanon

Military: **Active:**   56,000
            Defence budget: $631 million FMA (USA) 220 million (5.3% of total)
            % Middle East: 0.7

Index of Martial Potency: 4.50   **Rank:** 80

Political: **Type:** Republic   **Independence:** 22 November 1943
            Capital: Beirut
            Key individuals: President Rafiq Hariri, Hassan Nasrallah

Freedom in the World Rating: **PR:** 5   **CL:** 4   **Status:** PF

Failed State Index: 92.4   **Rank:** 28

Recent events

1975–90   Civil war ending with national reconciliation by the Ta'if Accord.
1978 and 1982   Invaded by Israel; the UN Interim Force in Lebanon (UNIFIL), a 200-strong body, has been operational in Lebanon since 1978.
2000   Israel withdrew from Southern Lebanon, having occupied the area since 1982.
2005   Prime Minister Rafiq Hariri assassinated; he had largely rebuilt the country since the end of the civil war.
            Syria withdrew its troops, their presence having been legitimised in the Ta'if Accord, from east Beirut and Bekaa Valley.
2006   34-day war with Israel resulting in $3.6 billion-worth of infrastructural damage.
2009   Parliamentary elections yield victory for pro-western March 14 Alliance under leadership of Saad Hariri.

Key issues

The continuing balance between the confessional communities; these comprise five recognised Islamic groups and eleven recognised Christian groups.
Lebanon's role as effectively a buffer state between Israel and Syria; unlike Jordan, this role has been complicated by a divided leadership.
The position of Hezbollah in the country; the organisation is allowed to retain its weapons while the Lebanese armed forces control only two-thirds of the country.
Hydropolitics: transfers of water from the River Litani, the only river wholly within Lebanon, voluntary or involuntary have been conjectured.
International boundary issues with Syria.
Production and transit of drugs.

Status

Like Jordan, Lebanon was carved out of Syria and, also like Jordan, it tends to be dominated by its powerful neighbours. An added complication in the case of Lebanon is the extraordinary confessional balance which was established at the state's foundation. The Muslim population is now said to be 59.7 per cent and the Christian 39 per cent. As a result of natural causes but also migration, the balance between the various communities is likely to change and this has constituted a continuing problem.

Lebanon is the smallest mainland state in the Middle East and, other than Cyprus, is the only Middle Eastern country with no desert landscape. Apart from Turkey and parts of Iran it is the only state which receives adequate rainfall. As a result, it can capitalise upon its only resources, agricultural products, which it exports throughout the Middle East. It has a small population, nearly 1 per cent of the Middle Eastern total, but a high literacy rate of 87.4 per cent. It achieves a GDP per capita which is over twice that of Jordan but, like Jordan, it has a wide range of trading links.

Owing to the civil war and the subsequent attack and occupation by Israel, large parts of the country were devastated and the new hope inspired by President Hariri was terminated by his assassination. What was once one of the most stable and prosperous countries in the Middle East has been torn apart and serious rebuilding: economic, social and political; is in its very early stages. It is therefore not surprising that Lebanon is rated twenty-eighth on the Failed State Index.

Politically, the country is badly split between pro- and anti-Syrian factions. Spearheading the former is the Hezbollah, a close ally of Iran and recently resurgent under the charismatic leadership of Hassan Nasrallah. In the 2006 war, Hezbollah militants surprised Israeli forces by the competence and doggedness of their military resistance and fought the mighty Israeli Defence Force (IDF) to a standstill. One of the key questions for the future political stability of Lebanon will be how to accommodate the increasing influence of Hezbollah within the existing framework of political institutions. In May 2008, violent clashes between Hezbollah and government forces erupted following a government decision to shut down a private telecommunications network controlled by Hezbollah and to close Beirut International Airport. Mediation by the Arab League resulted in the signing of the Doha Agreement in May 2008, which called for the formation of a power-sharing government of national unity to include significant Hezbollah representation. Though this agreement was able to resolve the immediate crisis, the stability of Lebanon's political system remains tenuous. However, the results of the 2009 parliamentary election appear to have helped stabilise Lebanon's politics, delivering a relatively clear-cut victory to the "pro-western", anti-Syrian March 14 Alliance.

LIBYA

Location:	Central in North Africa with a coastline along the entire central Mediterranean. (Map 43)

Area: 1,759,540 km²   **% Middle East:** 10.7
                    Comparison with Bahrain: 2,645.9×

Boundaries:	**Land: length:** 4,348 km	
	Contiguous countries: Niger 354km, Sudan 383 km, Tunisia 459 km, Algeria 982 km, Chad 1,055 km, Egypt 1,115 km	
	Maritime:	**Coastline:** 1,770 km
		Claims: territorial sea 12 nml with latitude 32°30' as the baseline exclusive fishing zone 62 nml
	Settlements:	Land: Algeria 1911, 1956 Chad 1902, 1994 Egypt 1925, 1926 Niger 1898, 1899, 1919, 1955 Sudan 1919, 1934 Tunisia 1910, 1956, 1970
		Maritime: Malta 1986 Tunisia 1988

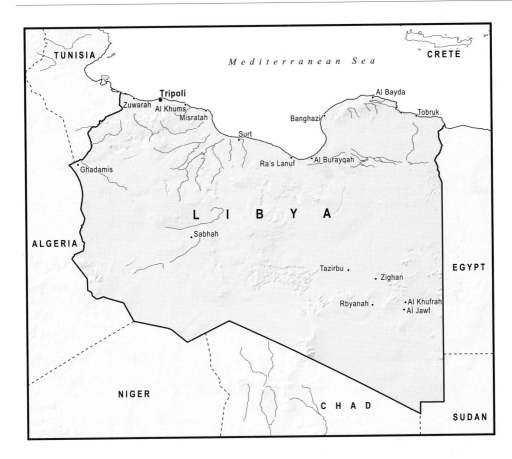

Map 43 Libya

Boundary Vulnerability Index: 3.5

Geography: **Topography:**   Coastal lowland in the north, mountains, plateaux and
plains to south in Sahara Desert.
Climate: Hot desert except for narrow Mediterranean coastal fringe.

Annual water withdrawal (domestic/industrial/agricultural): 4.27 cu km/yr
(14%/3%/83%)   **per capita:** 730 cu m/yr (2000)

Population: 6.0 million
% Middle East: 1.2   **Comparison with Bahrain:** 8.6×
Literacy: 82.6%
Ethnic groups: Berber and Arab 97%

Economy: **GDP:** $78.79 billion   **% Middle East:** 1.9
   Comparison with Bahrain: 3.2×   **GDP per capita:** $13,100
   Key resources: petroleum, natural gas
   Cropped land: 1.22%   **Irrigated land:** 4,700 km²

Human Development Index Rank: 56

Culture: **Languages:**   Arabic, Italian, English
   Religion: Muslim (Sunni) 97%
   Legal system: Based on Italian and French civil law and Islamic law.

Connectivity: **Telephones (fixed):** 483,000   **Mobiles:** 3.928 million
   Internet users: 232,000   **% of population:** 3.9
   Imports: Italy 18.9% Germany 7.9% China 7.5% Tunisia 6.3% France
       5.8% Turkey 5.2% USA 4.7% South Korea 4.3% UK 4.0%
   Exports: Italy 36.7% Germany 14.3% Spain 8.7% USA 6.1% France 5.6%
       Turkey 5.3%

Index of Globalisation: –   **Rank:** –

Military: **Active:**   76,000
   Defence budget: $650 million
   % Middle East: 0.7

Index of Martial Potency: 5.14   **Rank:** 63

Political: **Type:** Jamahiriya   **Independence:** 24 December 1951
       (State of the masses)
   Capital: Tripoli
   Key individuals: Muammar al-Qaddafi

Freedom in the World Rating: **PR:** 7   **CL:** 7   **Status:** NF

Failed State Index: 69.3   **Rank:** 115

Recent events

1969   Military coup by Colonel Muammar al-Qaddafi; new Constitutional Proclamation.

1973–87   Military operations conducted in northern Chad, ostensibly to gain access to minerals in the Aozou Strip.

1976   The Third Universal Theory of Qaddafi published in The Green Book.

1977   Adopted the Declaration of the Establishment of the People's Authority; the idea behind this was a unique form of direct democracy.

1989   Joined the Arab *Maghrib* Union; Libya implicated in the bombing of flight PA103 over Lockerbie, Scotland, in 1988, and UTA772 over Niger, in 1989; there remains grave doubt about the Libyan involvement in the Lockerbie disaster.

1992–2003 UN sanctions following the bombing of the flights.

2006   Full diplomatic relations with the USA resumed.

Key issues

Continuing influence of Qaddafi, the most significant and influential leader remaining in the
Middle East after the death of Assad, Khomeini and King Hussein and the execution of
Saddam Hussein.
Links between Africa and the Middle East.
Shortage of water.
International boundary issues with Tunisia and Malta and internationally over the Gulf of
Sirte.
Arab nationalism.

Status

Libya is a large country comprising mostly desert, located centrally on the north African coastline between Egypt and the *Maghrib* proper. Only Sudan, Algeria and Saudi Arabia are larger
in area among Middle Eastern states. Libya has six contiguous countries and there have been
disputes with all of them. However, these are now largely resolved other than over local issues
and Libya has a low Boundary Vulnerability Index. The key resources are petroleum and natural
gas and Libya enjoys the advantage of access to Western markets without the constraints of the
Suez Canal. Other than the main oil producers of the Arabian Peninsula, Libya has the largest
GDP per capita in the Middle East. This results from the size of the oil resources and the small
population in terms of which Libya is almost a micro-state.

As a predominantly arid country, Libya has developed the Great Man-made River Project, the
largest such water development scheme in the world, to transport water from the Fezzan, notably
the oasis of Kufra, to the coast. As a result, it now has a relatively large irrigated area, exclusively along the coastal zone, particularly around the Gulf of Sirte.

Links with Africa resulted in the establishment of a new African Union under the auspices of
Libya. However, people trafficking from sub-Saharan Africa and Asia is a serious problem and
the country is on the Tier Two Watch List.

Arguably, no country other than Iran and possibly Jordan has depended so much upon the
leadership of one man and has, in its foreign policy, reflected his style. The constant perception
of Libya has been its involvement with terrorism, particularly throughout the 1980s. Several of
the issues with which Libya has been associated have yet to be proved but there is strong
evidence that terrorism has been generated and there have been cross-boundary transfers of
arms. With time, Qaddafi appears to have moderated his views and, in 2006, the USA rescinded
Libya's designation as a state sponsor of terrorism. Full diplomatic relations with the USA and
Britain were restored after Qaddafi agreed to terminate ongoing WMD (mainly chemical)
programmes.

MAURITANIA

Location:	North-west Africa, between Morocco/Algeria and Senegal/Mali with an Atlantic coastline. (Map 44)

Area: 1,030,700 km²   **% Middle East:** 6.3
Comparison with Bahrain: 1,549.9×

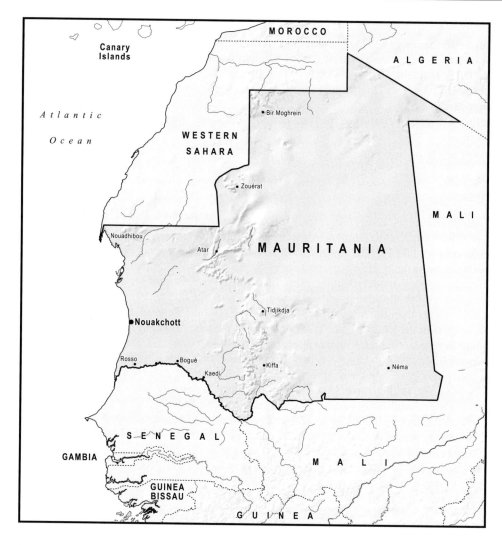

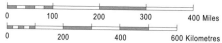

Map 44 Mauritania

Boundaries: **Land: length:** 5,074 km
Contiguous countries: Algeria 463 km, Senegal 813 km, Western
Sahara 1,561 km, Mali 2,237 km
Maritime: **Coastline:** 754 km
Claims: territorial sea 12 nml EEZ 200 nml
contiguous zone 24 nml
continental shelf to 200 nml or continental margin

Settlements: Land: Algeria 1909, 1963, 1985 Mali 1944, 1963
Senegal 1904, 1905, 1933 Western Sahara 1900, 1912, 1956
Maritime: Morocco 1976

Boundary Vulnerability Index: 7.5

Geography: **Topography:** Coastal plain, with mountains and desert to the east.
Climate: Hot and arid desert except for coastal maritime influence.

Annual water withdrawal (domestic/industrial/agricultural): 1.7 cu km/yr
(9%/3%/88%) **per capita:** 554 cu m/yr (2000)

Population: 3.3 million
% Middle East: 0.6 **Comparison with Bahrain:** 4.6×
Literacy: 51.2%
Ethnic groups: Mixed Moor/Black 40% Moor 30% Black 30%

Economy: **GDP:** $5.818 billion **% Middle East:** 0.1
Comparison with Bahrain: 24% **GDP per capita:** $1,800
Key resources: iron ore, fish
Cropped land: 0.21% **Irrigated land:** 490 km²

Human Development Index Rank: 137
Culture: **Languages:** Arabic (official), Pulaar, Soninke, French, Hassawiya, Wolof
Religion: Muslim 100%
Legal system: Combination of Islamic law and French civil law

Connectivity: **Telephones (fixed):** 35,000 **Mobiles:** 1.06 million
Internet users: 100,000 **% of population:** 3.0
Imports: France 11.9% China 8.1% Belgium 6.8% USA 6.7% Italy 5.9%
Spain 5.7% Brazil 5.5%
Exports: China 26.1% Italy 11.7% France 10.5% Spain 6.9%
Belgium 6.8% Japan 5.4% Cote d'Ivoire 4.6%

Index of Globalisation: – **Rank:** –

Military: **Active:** 15,870
Defence budget: $18.6 million
% Middle East: 0.02

Index of Martial Potency: 2.69 **Rank:** 129

Political: **Type:** Republic **Independence:** 28 November 1960
Capital: Nouakchott
Key individuals: –

Freedom in the World Rating: **PR:** 6 **CL:** 4 **Status:** PF

Failed State Index: 86.7 **Rank:** 45

Recent events

1976   Annexed southern third of Western Sahara after the Madrid Agreement in which Spain ceded the territory of Western Sahara.
1978   Military coup.
1979   Relinquished its part of Western Sahara after Polisario raids; Algiers agreement with Polisario.
1984   Coup by Maaouya Taya.
2001   Legislative elections judged free and open.
2005   Military coup; ruled by a Military Council for Justice and Democracy.
2006–7 Parliamentary and presidential elections.

Key issues

Arab World–Africa boundary; tensions between black population and Arab Berber communities.
High level of poverty: over 50 per cent of the population depends upon agriculture and livestock.
Key resources: iron ore and fish; 2006 oil exports began.
Extreme water shortage; River Senegal along the southern boundary is the only perennial river.
Refugees: especially from Mali; plans to demarcate international boundary.
Source and destination for slavery-related practices; people trafficking.

Status

Mauritania is located at the interface between the Arab world and Africa. To the north are Western Sahara, Morocco and Algeria and to the east and south Mali and Senegal. It is the sixth largest state in the Middle East, slightly larger than Egypt. In contrast, apart from Oman, Kuwait, Qatar, Cyprus, Bahrain and Western Sahara, it has the smallest population. As a result, it has by far the lowest density of population in any Middle Eastern country. There are four contiguous states and boundary settlements have been reached with all of them although there are minor disputes about grazing rights with Senegal. A maritime boundary agreement was made with Morocco in 1976 but that followed the annexation of Western Sahara and relates to what was then the land boundary. Therefore, it appears on the map but is essentially invalid, the only example of such a maritime boundary in the world.

Apart from the coastal plain, Mauritania comprises hot and arid desert with extremely limited opportunities for agriculture or irrigation. Aside from the future Palestine, it has by far the lowest GDP in the Middle East although, owing to its small population, its per capita GDP exceeds those of Afghanistan, the future Palestine, Yemen and, marginally, Sudan. Apart from Afghanistan and Iraq, it has the lowest percentage of population using the Internet among Middle Eastern countries. Its defence budget is by far the smallest accounting for only 0.02 per cent of the total Middle Eastern defence budget. At 51.2 per cent the literacy rate of the population is well above that of Afghanistan and marginally ahead of that of Yemen but otherwise the lowest in the Middle East; it is 10 per cent below that of Sudan. The ethnicity of the population illustrates clearly the interface between Arab and African. The ethnic groups are listed as mixed Moor/Black 40 per cent, Moor 30 per cent, Black 30 per cent.

By any standards, Mauritania is a very poor country but there are development prospects hingeing on oil and iron ore. A barter economy still exists in parts of the country where there are also slavery-related practices. Relations with Africa are more concerned with trafficking than legitimate trade as shown by the list of import and export partners. The country is placed on the Tier Two Watch List. Mauritania is ranked forty-five on the Failed State Index.

MOROCCO

Location:	Westernmost *Maghrib* state in the north-west corner of Africa with coast-lines on the Mediterranean, the Strait of Gibraltar and the Atlantic Ocean. (Map 45)

Area: 446,550 km²   **% Middle East:** 2.7
Comparison with Bahrain: 671.5×

Boundaries: **Land: length:** 2,017.9 km
Contiguous countries: Spain (exclaves of Ceuta and Melilla) 15.9 km, Western Sahara 443 km, Algeria 1,559 km
     Maritime: **Coastline:** 1,835 km
              Claims: territorial sea 12 nml EEZ 200 nml
                      contiguous zone 24 nml
                      continental shelf: 200 m depth or depth of exploitation
     Settlements: Land: Algeria 1845, 1972 Spain no agreement
                    Western Sahara 1904, 1912

Boundary Vulnerability Index: 7.5

Geography: **Topography:**   Hills and lowlands in north rising to the Atlas Mountains in centre and desert plains and plateaux to the south.
              Climate: Mediterranean on coast, modified inland by altitude. Hot and arid to south.

Annual water withdrawal (domestic/industrial/agricultural): 12.6 cu km/yr
       (10%/3%/87%)   **per capita:** 400 cu m/yr (2000)

Population: 33.8 million
              % Middle East: 6.7   **Comparison with Bahrain:** 48.3×
              Literacy: **52.3%**
              Ethnic groups: Arab Berber 99.1%

Economy: **GDP:** $127.0 billion   **% Middle East:** 3.0
              Comparison with Bahrain: 5.2×   **GDP per capita:** $3,800
              Key resources: phosphates
              Cropped land: 21.0%   **Irrigated land:** 14,450 km²

Human Development Index Rank: 126

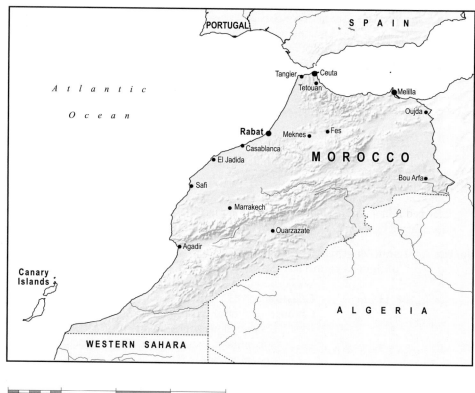

Map 45 Morocco

Culture: **Languages:** Arabic (official), Berber, French
Religion: Muslim 98.7% Christian 1.1%
Legal system: Based upon Islamic law and French and Spanish civil law.

Connectivity: **Telephones (fixed):** 1.266 million **Mobiles:** 16.005 million
Internet users: 6.1 million **% of population:** 18.0
Imports: France 17.51% Spain 13.9% Saudi Arabia 6.9% China 6.9%
Italy 6.3% Germany 6.0%
Exports: Spain 20.6% France 20.5% UK 4.8% Italy 4.7% India 4.0%

Index of Globalisation: 52.93 **Rank:** 70

Military: **Active:** 195,800
Defence budget: $2.47 billion FMA (USA) 12.5 million (0.3% of total)
% Middle East: 2.5

Index of Martial Potency: 5.48 **Rank:** 47

Political: **Type:** Constitutional Monarchy   **Independence:** 2 March 1956
                      Capital: Rabat
                      Key individuals: King Hassan II

Freedom in the World Rating: **PR:** 5   **CL:** 4   **Status:** PF

Failed State Index: 76   **Rank:** 86

Recent events

1956   Upon independence from France, most Spanish possessions were given to the new
        state.
1961   King Hassan II came to the throne.
1962   Claimed mineral rich Tindouf area of Algeria.
1969   Abandoned claim to Mauritania.
1974   Morocco and Mauritania made a secret agreement to divide Spanish Sahara.
1975   The ICJ ruled in favour of self-determination for Spanish Sahara; the Green March led
        by the King of 350,000 unarmed civilians to reinforce the claim to Spanish Sahara.
1976   Spanish troops left Spanish Sahara.
1978   Mauritania renounced claims to Spanish Sahara.
1989   International boundary treaty with Mauritania ratified.
1997   First meeting of the new Bicameral Legislature.
1999   King Hassan II died to be succeeded by his son, King Mohamed VI.

Key issues

Spanish exclaves: Ceuta, Melilla and Penon de Velez de la Gomera, the islands of Penon de
        Alhucemas and Islas Chafarinas and surrounding waters; Morocco rejected Spain's unilat-
        eral designation of a maritime boundary comprising a median line from the Canary
        Islands; 2002 dispute with Spain over the Islet of Perejil (Leila in Morocco) occupied by
        Moroccan troops.
The Strait of Gibraltar as a choke point and also the status of the Spanish exclaves and Gibraltar
        itself.
Western Sahara: virtually annexed in the late 1970s; the status of the territory remains
        unresolved.
Key starting point for illegal immigration to Spain and the EU.
One of the world's largest producers of illicit hashish and a transit point for cocaine from South
        America to Western Europe.

Status

Morocco is strategically located on the Strait of Gibraltar with both Mediterranean and Atlantic coastlines. Contiguous countries are Algeria and Western Sahara with both of which there
have been major disagreements. Legally, through its various exclaves, Spain is also contiguous and its presence, strongly disputed by Morocco, raises a wide variety of boundary issues.
It is, in fact, a unique situation in which a land boundary with another state needs to be
followed geographically by a maritime boundary with that same state. The topography of

Morocco is typical of the *Maghrib* with a coastal plain to the north being succeeded by mountains and then desert and a climate ranging from Mediterranean in the north to hot and arid desert in the south.

Its mixed Arab Berber population has a literacy rate of 52.3 per cent well below that of the other major *Maghrib* states (Tunisia 74.3 per cent and Algeria 69.9 per cent). The GDP per capita is also significantly lower than that of those states, being barely more than half that of Tunisia and well under half that of Algeria. The key resource is phosphates of which it has globally important reserves. Agriculture remains important and Morocco has approximately three times the irrigated areas of Algeria and Tunisia. Its trading pattern is more diverse than that of Tunisia but less cosmopolitan than that of Algeria. Another variable is the defence budget: Morocco's is about two-thirds that of Algeria but five times that of Tunisia. On the Freedom in the World Rating it ranks well above both Algeria and Tunisia. Given its proximity to the EU, Morocco strives for Western standards but has a low GDP per capita and remains a focus of illegal drugs trading.

Morocco has experienced serious problems with terrorism, including a series of bombings in Casablanca in 2003 that left forty-five people dead. A subsequent government crackdown on Islamist groups has resulted in more than thirty groups being dismantled, but Morocco still remains a prime recruiting ground for Islamist groups seeking foot soldiers for Iraq.

OMAN

Location:	South-east end of the Arabian Peninsula with coastlines on the Gulf of Oman and the Arabian Sea and exclaves within the UAE and on the south side of the Strait of Hormuz. (Map 46)

Area: 212,460 km²   **% Middle East:** 1.3
Comparison with Bahrain: 319.5×

Boundaries: **Land: length:** 1,374 km
Contiguous countries: Yemen 288 km, UAE 410 km, Saudi Arabia 676 km
Maritime: **Coastline:** 2,092 km
Claims: territorial sea 12 nml EEZ 200 nml
contiguous zone 24 nml
Settlements: Land: Saudi Arabia 1920, 1990 UAE no agreements but reported settlement in 2003 Yemen 1992
Maritime: Iran 1947 Pakistan 2000

Boundary Vulnerability Index: 1.5

Geography: **Topography:**   North-south: a coastal plain, a mountain range, a central desert region, a southern mountain range and a small coastal plain.
Climate: Hot and arid modified by altitude and maritime influence.

Annual water withdrawal (domestic/industrial/agricultural): 1.36 cu km/yr
(7%/2%/90%)   **per capita:** 529 cu m/yr (2000)

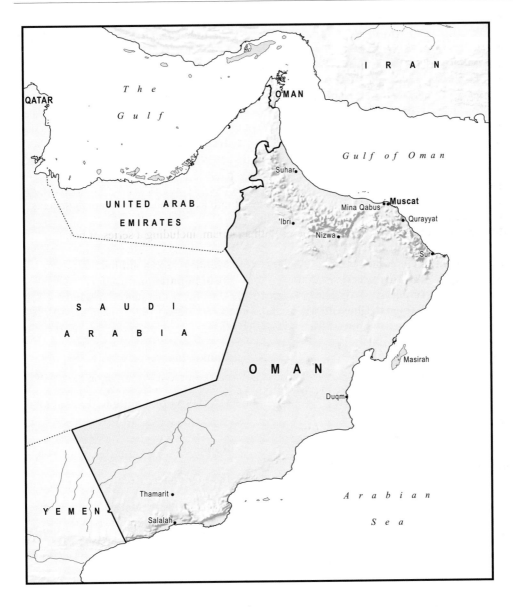

Map 46 Oman

Population: 3.2 million
  % Middle East: 0.6   **Comparison with Bahrain:** 4.6×
  Literacy: 81.4%
  Ethnic groups: Arab, Baluchi, South Asian, African

Economy: **GDP:** $61.21 billion   **% Middle East:** 1.5
  Comparison with Bahrain: 2.5×   **GDP per capita:** $19,100
  Key resources: petroleum
  Cropped land: 0.26%   **Irrigated land:** 720 km²

Human Development Index Rank: 58

Culture: **Languages:**   Arabic (official), English, Baluchi, Urdu, Indian dialects
  Religion: Ibadhi Muslim 75%
  Legal system: Based on English common law and Islamic law

Connectivity: **Telephones (fixed):** 278,300   **Mobiles:** 1.818 million
  Internet users: 19,000   **% of population:** 10.0
  Imports: UAE 22.4% Japan 16.4% USA 8.1% Germany 5.5% India 4.3%
  Exports: China 23.6% South Korea 17.9% Japan 10.9% Thailand 10.7%
    South Africa 7.7% UAE 6.3%

Index of Globalisation: 51.67   **Rank:** 76

Military: **Active:**   42,600
  Defence budget: $3.23 billion FMA (USA) 14.0 million (0.3% of total)
  % Middle East: 3.3

Index of Martial Potency: 5.91   **Rank:** 36

Political: **Type:** Monarchy   **Independence:** 1650
  Capital: Muscat
  Key individuals: Sultan Qaboos

Freedom in the World Rating: **PR:** 6   **CL:** 5   **Status:** NF

Failed State Index: 45.5   **Rank:** 146

Recent events

1970   Sultan Qaboos deposed his father as ruler and took on the role of Prime Minister.
1972   Qaboos began a programme of relatively rapid modernisation which included the incorporation of the Omanis from Zanzibar, Oman having been the only non-European country to have an African colony.
1976   Dispute with South Yemen began.
1990   International boundary settlement with Saudi Arabia.
1992   International boundary settlement with Yemen.

2003   International boundary agreement reportedly signed and ratified with the UAE for the entire boundary including that in the Musandam Peninsula and the Al Mahdah exclaves.

Key issues

Rate and character of modernisation.
Relative detachment from the other countries of the GCC.
Location on the Strait of Hormuz.
People trafficking.

Status

Located at the eastern end of the Arabian Peninsula, Oman is insulated from the other countries by mountains and desert. It has long coastlines on the Gulf of Oman and the Arabian Sea and has a strong maritime tradition. Its independence of action is illustrated by its continuing close links with Iran and by its operations with the US Rapid Deployment Force which has a permanent base on Masirah Island. Crucially, Oman also has an exclave which forms the southern shore of the Strait of Hormuz, at the tip of the Musandam Peninsula.

Topographically, Oman appears to comprise three countries in one: a northern developed coastal plain stretching along the southern shore of the Gulf of Oman with Muscat as its focus; a central area comprising a major mountain chain and attendant plateaux with its centre Nizwa; and a southern coastal plain, the Dhofar, which focuses upon Salalah and is cut off from the Nizwa area by mountains and an extensive track of hot desert, in part the Rub al-Khali.

In terms of population, it is a micro-state, but its GDP per capita is exceeded only by those of the Qatar, Kuwait, the UAE, Bahrain, Israel, Cyprus and Saudi Arabia. It is marginally below that of Saudi Arabia. The population comprises a number of ethnic groups: Arab, Baluchi, South Asian and African, but has a relatively high literacy rate. Languages include a range of Indian dialects and there is a variety of religions although Ibadhi Muslims comprise 75 per cent of the population.

The key resource is petroleum although Oman is considerably less well endowed than any other member of the GCC except Bahrain. Oman ranks high in Martial Potency at thirty-sixth in the world with one of the larger defence budgets in the Middle East. Indeed, it is only fractionally smaller than that of Egypt. The armed forces are relatively small but well balanced and extremely well trained, largely by the UK, with high quality equipment. Oman receives military assistance from the USA and is a significant actor in Middle Eastern affairs as a result of the thoughtful diplomacy of Sultan Qaboos. However, Oman remains a destination country for people trafficking largely from the Indian subcontinent to involuntary servitude. It is placed on the Tier Two Watch List.

With a ranking of 146 on the Failed State Index, Oman is thought to be the most stable and least prone to state failure of all Middle Eastern countries.

QATAR

Location: Small promontory located centrally on the south coastline of the Gulf.
                     (Map 47)

Area: 11,437 km²   **% Middle East:** 0.07
                              Comparison with Bahrain: 17.2×

Boundaries: **Land: length:** 60 km
                     Contiguous countries: Saudi Arabia 60 km
                     Maritime: **Coastline:** 563 km
                                         Claims: territorial sea 12 nml
                                                     EEZ as determined by bilateral agreement or the
                                                     median line
                                                     contiguous zone 24 nml
                     Settlements: Land: Saudi Arabia 1914, 1965, 1993
                                         Maritime: Bahrain 2001, Iran 1969, UAE 1969
                                         UAE maritime boundary affected by the UAE–Saudi
                                         Arabia agreement on the Saudi Arabian corridor

Boundary Vulnerability Index: 4

Geography: **Topography:**   Low plateau
                     Climate: Hot and arid desert with minor maritime moderation.

Annual water withdrawal (domestic/industrial/agricultural): 0.29 cu km/yr
                     (24%/3%/72%)   **per capita:** 358 cu m/yr (2000)

Population: 907,000
                     % Middle East: 0.2   **Comparison with Bahrain:** 1.3×
                     Literacy: 89.0%
                     Ethnic groups: Arab 40% Indian 18% Pakistani 18% Iranian 10%

Economy: **GDP:** $57.69 billion   **% Middle East:** 1.4
                     Comparison with Bahrain: 1.5×   **GDP per capita:** $75,900
                     Key resources: natural gas, petroleum
                     Cropped land: 1.91%   **Irrigated land:** 130 km²

Human Development Index Rank: 35

Culture: **Languages:**   Arabic (official), English
                     Religion: Muslim 77.5% Christian 8.5%
                     Legal system: Based upon Islamic and civil law codes.

Connectivity: **Telephones (fixed):** 228,300   **Mobiles:** 919,800
                     Internet users: 290,000   **% of population:** 32.0
                     Imports: France 13.3% Japan 10.2% USA 9.3% Italy 8.9% Germany 7.7%
                                  UK 6.2% Saudi Arabia 5.7% South Korea 4.7%
                     Exports: Japan 40.2% South Korea 16.4% Singapore 6.5% Thailand 4.1%

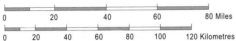

Map 47 Qatar

Index of Globalisation: –   **Rank:** –

Military: **Active:**   11,800
                 Defence budget: $2.33 billion
                 % Middle East: 2.4

Index of Martial Potency: –   **Rank:** –

Political: **Type:** Emirate   **Independence:** 3 September 1971
                 Capital: Doha
                 Key individuals: al-Thani family

Freedom in the World Rating: **PR:** 6   **CL:** 5   **Status:** NF

Failed State Index: 53.6   **Rank:** 137

Recent events

1971   Discussions on terms of Federation with Bahrain and the seven Sheikhdoms which comprised the Trucial Oman having failed, Qatar became independent.
1986   Commencement of a long-term boundary dispute with Bahrain.
1995   Coup by the current Emir, Hamad bin Khalifa al-Thani.
2001   Boundary disputes with Bahrain and Saudi Arabia settled.
2003   New, democratic constitution endorsed in popular referendum.

Key issues

Overwhelming importance of petroleum and particularly gas in the economy.
Establishment and development of the Saudi maritime corridor which separates Qatar from the
    UAE.
Ethnic make-up of the population.
Water shortage.
People trafficking.

Status

Qatar comprises a small peninsula on the Gulf coast of the Arabian Peninsula. It is located near Bahrain, midway between the Strait of Hormuz and Kuwait and is therefore centrally placed with regard to petroleum-related turbulence. Among Middle Eastern countries, only Bahrain, the future Palestinian state, Cyprus and Lebanon occupy a smaller area, while only Bahrain, Cyprus and Western Sahara have a lower population. It is in every sense a micro-state but it has the crucial asset of petroleum. Oil provides approximately 80 per cent of export earnings and Qatar's proved reserves of natural gas are the third largest in the world. As a result, it has a GDP per capita which is the highest in the Middle East, almost 1.4 times larger than those of Kuwait and the UAE. Its defence budget is similar to that of Morocco, a far larger state in every respect.

The population has a high literacy rate but is only 40 per cent Arab; 36 per cent of the population is listed as Indian or Pakistani and 10 per cent as Iranian. Given its obvious wealth, the security of Qatar depends very much upon a balancing act internally and externally. It is one of the few states which has maintained close relations with both the USA and Iran, to which it is linked by a water pipeline. Indeed the USA used the military base at al-Udeid as the command centre for operations during Operation Iraqi Freedom, and maintains the forward headquarters of the Coalition Command and US Central Command (CENTCOM) at Camp as-Saliyah.

Plans to move towards a more democratic political system appeared to be progressing with the passage, in 2003, of a new constitution that envisaged a new national assembly with two-thirds of its members to be elected. However, little movement has been made to date to implement the provisions of the new constitution.

SAUDI ARABIA

Location: Occupies the major part of the Arabian Peninsula with coastline on the Red Sea and the Gulf. (Map 48)

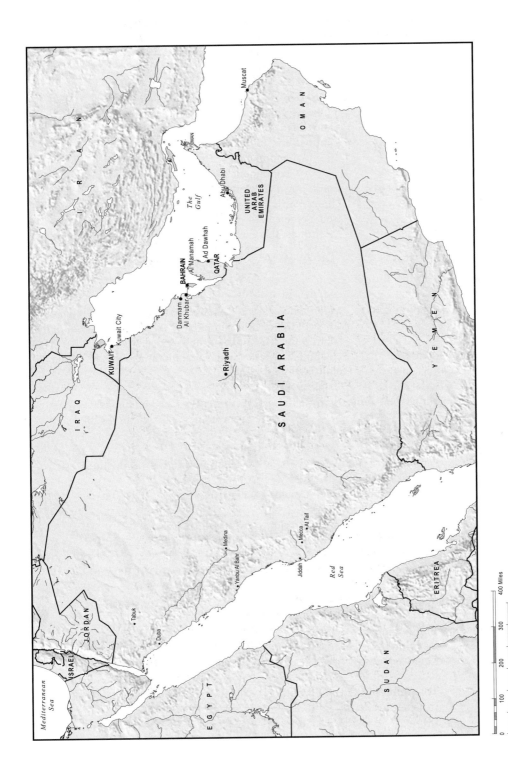

Map 48 Saudi Arabia

Area: 2,149,690 km²   **% Middle East:** 13.1
                    Comparison with Bahrain: 3,232.6×

Boundaries: **Land: length:** 4,431 km
          Contiguous countries: Qatar 60 km, Kuwait 222 km, UAE 457 km,
                    Oman 676 km, Jordan 744 km, Iraq 814 km,
                    Yemen 1,458 km
          Maritime: **Coastline:** 2,640 km
                    Claims: territorial sea 12 nml
                         continental shelf not specified
                         contiguous zone 18 nml
          Settlements: Land: Iraq 1922 Jordan 1925, 1965 Kuwait 1922, 1961
                    Oman 1920, 1990 Qatar 1914 1965 1993 UAE to be final-
                    ised Yemen 1914, 1995, 2000
                    Maritime: Bahrain 1958 Iran 1968 Kuwait 2000 Yemen
                    2000

Boundary Vulnerability Index: 2.5

Geography: **Topography:**   Mountainous along Red Sea coast with a low-lying area
                    adjacent to the Gulf and largely hot desert between. In the
                    south is the Rub al Khali sand sea.
          Climate: Hot and arid desert with some modification by altitude and
                    maritime effect.

Annual water withdrawal (domestic/industrial/agricultural): 17.33 cu km/yr
          (10%/1%/89%)   **per capita:** 705 cu m/yr (2000)

Population: 27.6 million (non-nationals 5.5 million)
          % Middle East: 5.5   **Comparison with Bahrain:** 39.4×
          Literacy: 78.8%
          Ethnic groups: Arab 90% Afro-Asian 10%

Economy: **GDP:** $572.2 billion   **% Middle East:** 13.7
          Comparison with Bahrain: 23.3× **GDP per capita:** $20,700
          Key resources: petroleum, natural gas
          Cropped land: 1.76%   **Irrigated land:** 16,200 km²

Human Development Index Rank: 61

Culture: **Languages:**   Arabic
          Religion: Muslim 100%
          Legal system: Based upon *Shari'a* law

Connectivity: **Telephones (fixed):** 4.5 million   **Mobiles:** 19.663 million
          Internet users: 4.7 million   **% of population:** 17.0
          Imports: USA 12.2% Germany 9.1% China 7.9% Japan 7.3% UK 4.8%
                    Italy 4.8% South Korea 4.1
          Exports: Japan 17.7% USA 15.8% South Korea 9.0% China 7.2%
                    Taiwan 4.6% Singapore 4.4%

Index of Globalisation: 53.69   **Rank:** 68

Military: **Active:**   223,500
                    Defence budget: $33.33 billion
                    % Middle East: 34.4

Index of Martial Potency: 6.76   **Rank:** 17

Political: **Type:** Monarchy   **Independence:** 23 September 1932
                    Capital: Riyadh
                    Key individuals: King Abdul al-Aziz and the al-Saud dynasty

Freedom in the World Rating: **PR:** 7   **CL:** 6   **Status:** NF

Failed State Index: 76.5   **Rank:** 83

Recent events

1973   Arab–Israeli War, Saudi Arabia embargoed oil shipments to the USA and cut oil exports by 10 per cent, thereby paving the way for the OPEC 1973–4 price rises.
1979   Seizure of the Grand Mosque in Mecca by insurgents; the year of the Iranian Revolution.
1990   Iraq's invasion of Kuwait. Saudi Arabia requested US support.
2003   All operational US troops left Saudi Arabia; major terrorist attacks occurred.
2005   Election for half the members of the 179 municipal councils.

Key issues

Islam, Wahhabism and Muslim conservatism.
Petroleum: Saudi Arabia has 21.3 per cent of the world's oil reserves.
Water shortage.
Maritime access.
Population: rapid increase. Unemployment and number of foreign workers.
People trafficking.

Status

Saudi Arabia occupies the greater part of the Arabian Peninsula but has no direct access to either the Gulf of Oman or the Arabian Sea. It is the spiritual heart of Islam, its birthplace and the home of the two holy shrines at Mecca and Medina. Wahhabism, which calls for the purification of Islam and a return to the fundamentals of the faith, is dominant in Saudi Arabia. At an early date, the Wahhabi formed a political alliance with the al-Saud family and when, under the inspired leadership of Ibn Saud, the holy cities of Mecca and Medina were conquered and the state was unified (1932), the partnership became central in Saudi Arabian affairs.

In 1938 came the momentous discovery of oil in the eastern province, al-Hasa. Saudi Arabia has three megafields and the world's largest oil reserves, 21.3 per cent of the proved total. This fact alone makes it a key actor on the world stage but also makes it a potential target for aggression. Having no direct access to the world's ocean, oil exports from Saudi Arabia have to transit the Strait of Hormuz, the Suez Canal or the Strait of Bab al Mandab. Saudi Arabia has been able to exercise effective diplomacy in various inter-Arab disputes but has also had to spend vast sums on defence and has still been involved in most of the conflicts within the region. Despite its huge defence expenditure, Saudi Arabia invariably requests US support in times of turbulence and this has led to a fundamental national debate between those who favour increased Islamic morality and those who support greater democracy.

Within the Middle East, Saudi Arabia is exceeded in area only by Algeria and Sudan but in terms of population is only the seventh largest state, marginally larger than Iraq. Apart from the two coastal plains, it comprises mountainous areas but largely hot desert including the Rub al-Khali, the largest sand sea in the world.

Given its position in the world oil market, Saudi Arabia has a relatively low GDP per capita, only the seventh largest in the region. Its export and import partners show a relatively narrow trading spectrum although there is a discernable move towards increased oil exports to the Far East at the expense of Western Europe and the USA.

Saudi Arabia has by far the largest defence budget in the Middle East. At $33.33 billion it is three times larger than the next largest, that of Turkey, and accounts for 34.4 per cent of the total Middle Eastern defence budget. There have been terrorist incidents, but by and large Saudi Arabia has remained a relative island of calm in the troubled area of the Gulf. However, internal problems are likely to be enhanced by the increasing number of educated, unemployed young people in a country in which 20 per cent of the population is non-national, largely brought in as labour. There are approximately 5.5 million foreign workers. The growing alienation of the country's repressed Shi'a community, which lives predominantly in the oil-rich areas, especially al-Hasa, is also likely to create problems in the future. Perhaps more than any other country, Saudi Arabia is directly threatened by the emergence of the so-called "Shi'a Crescent".

Saudi Arabia is also a focus for people trafficking, for involuntary servitude and commercial sexual exploitation. The major sources include south and south-eastern Asia, Nigeria, Yemen, Afghanistan, Somalia, Mali and Sudan. Saudi Arabia is listed in Tier Three of the Watch List.

SUDAN

Location: Largest state in Africa, located centrally in the continent, comprising largely the upper drainage basin of the River Nile. A coastline on the Red Sea. (Map 49)

Area: 2,505,810 km² **% Middle East:** 15.3
Comparison with Bahrain: 3,768.1×

Boundaries: **Land: length:** 7,687 km
Contiguous countries: Kenya 232 km, Libya 383 km, Uganda 435 km, Eritrea 605 km, Democratic Republic of the Congo 628 km, Central African Republic 1,165 km, Egypt 1,273 km, Chad 1,360 km, Ethiopia 1,606 km
Maritime: **Coastline:** 853 km
Claims: territorial sea 12 nml
continental shelf to depth of 200 m or depth of exploitation
contiguous zone 18 nml

Map 49 Sudan

Settlements: Land: Central African Republic 1924 Chad 1919 Democratic Republic of the Congo 1894, 1924 Egypt 1899 Eritrea 1992 Ethiopia 1891, 1898, 1899, 1901, 1902, 1903, 1916, 1972 Kenya 1914, Libya 1919, 1934 Uganda no agreements

Boundary Vulnerability Index: 9

Geography: **Topography:**   The north is largely desert with mountains on the Red Sea coast. The south is low plain, drainage basin of the River Nile.

Climate: Hot and arid in the north, hot, wet and tropical in the south.

Annual water withdrawal (domestic/industrial/agricultural): 37.32 cu km/yr
(3%/1%/97%)   **per capita:** 1,030 cu m/yr (2000)

Population: 39.4 million
% Middle East: 7.8   **Comparison with Bahrain:** 56.3x
Literacy: 61.1%
Ethnic groups: Black 52% Arab 39% Beja 6%

Economy: **GDP:** $107.8 billion   **% Middle East:** 2.6
Comparison with Bahrain: 4.4×   **GDP per capita:** $2,500
Key resources: petroleum, cotton
Cropped land: 6.95%   **Irrigated land:** 18,630 km²

Human Development Index Rank: 147

Culture: **Languages:**   Arabic (official), Nubian, Ta Bedawie, Nilotic, Nilo-Hamitic, Sudanic, English
Religion: Muslim (Sunni 70%) Christian 5% Indigenous beliefs 25%
Legal system: Based on English common law and Islamic law

Connectivity: **Telephones (fixed):** 637,000   **Mobiles:** 4.683 million
Internet users: 3.5 million   **% of population:** 8.9
Imports: China 18.2% Saudi Arabia 9.2% UAE 5.8% Egypt 5.3%
Germany 5.2% India 4.6% France 4.1%
Exports: Japan 48.0% China 31.0% South Korea 3.8%

Index of Globalisation: –   **Rank:** –

Military: **Active:**   109,300
Defence budget: $579 million
% Middle East: 0.6

Index of Martial Potency: 4.59   **Rank:** 78

Political: **Type:** Government of National Unity   **Independence:** 1 January 1956
Capital: Khartoum
Key individuals: President Gaafar Nimeri

Freedom in the World Rating: **PR:** 7   **CL:** 7   **Status:** NF

Failed State Index: 113.7   **Rank:** 1

Recent events

1969   Coup by Gaafar Nimeri; he was overthrown for three days after he made an agreement to join Egypt, Libya and Sudan in a single Federal State which never materialised.

1972–83 The south was autonomous.

1983   The southern autonomy ended when Nimeri re-divided the region into three and implemented *Shari'a* law.

1989   Military coup. Omar Bashir set up the Revolutionary Command Council for National Salvation.

1991   Sudan aligned with Iraq in the Gulf War.

1994   Sudan allegedly implicated in the World Trade Center bombing.

1997   China granted rights to exploit the largest oilfield.

1999   Inauguration of 1,610 km oil pipeline to Port Sudan, funded by China, Malaysia and Canada.

2003   Darfur region conflict began.

Key issues

Civil wars.
Political and economic instability.
Northern domination of south.
Hydropolitics of the River Nile.
Refugees and people trafficking.
International boundaries.

Status

In area, Sudan is the largest state in Africa and the largest state in the Middle East. The only other states which are over 2 million km^2 in area are Algeria and Saudi Arabia. Sudan accounts for over 15 per cent of the area of the Middle East. It is located centrally in Africa and comprises predominantly the upper basin of the River Nile and its tributaries. While geographically it might be counted as Middle Eastern, Sudan abuts on to only two other Middle Eastern states, Libya and Egypt. There are nine contiguous countries, a larger number and therefore a greater potential source of conflict than that found in any other Middle Eastern country. The only outlet to the global ocean is through a relatively short length of coastline on the Red Sea, part of which is disputed with Egypt. Given its location, relations with Egypt are crucial.

The dominating issue in Sudan is however the split between the Muslim north and the Christian and animist south, where there has been crisis since 1983. A succession of civilian and military governments has failed to end the conflict and has in many ways exacerbated the situation. Since 2003, the Darfur region has witnessed continuous fighting with the Sudanese Liberation Movement (SLM) and the Justice and Equality Movement (JEM) joining forces against the government. The government used ethnic Arab militias including the dreaded Janjaweed. The situation has resulted in mass migrations of people, starvation, the inclusion of children in the fighting and deaths on a vast scale. The second civil war resulted in the displacement of four million people and more than two million deaths over a period of two decades. This was apparently settled by the final north-south Comprehensive Peace Agreement (CPA) of 2005 which granted the southern rebels autonomy for six years following which a referendum will be held on independence.

However, prior to this, the Darfur conflict had begun in which some two million people have been displaced and up to 400,000 people killed. At the same time, instability has spread to eastern Chad and the Sudanese incursions have brought conflict to the Central African Republic. Humanitarian assistance has been hampered by the poor infrastructure and the lack of government support. Furthermore, Sudan abuts on to the Democratic Republic of the Congo, itself the scene of continuing conflict with its neighbours. Relations with Egypt, Kenya and Uganda continue to be poor. Nonetheless, the Agreement for the Full Utilisation of the Nile Waters, signed in 1959 with Egypt, remains in operation. This ratified the Nile Water Agreement of 1929 which gave Egypt 48 bcm and Sudan 4 bcm per year. The new agreement meant that additional water available from the formation of Lake Nasser was shared between Sudan and Egypt in a ratio of approximately 2:1.

The population of Sudan is out of proportion with the size of the country and accounts for only 7.8 per cent of the Middle Eastern total. The main ethnic groups are Black, Arab and Beja and the north-south conflict is reflected in these ethnic groupings. As a result of political instability, civil war, hyperinflation and reduction in remittances from abroad, the economy is in a parlous position. Sudan is potentially a wealthy country but has a GDP per capita of only $2,500, a figure below that of every other Middle Eastern state except Yemen, Mauritania, Palestine (Gaza Strip and the West Bank) and Afghanistan. The languages also reflect the national divide, comprising Arabic and a wide range of African languages. Among the religions, indigenous beliefs account for 25 per cent of the population.

Apart from the large-scale movements of arms, militias and refugees across boundaries, there are two outstanding international boundary disputes. One is to the west of Lake Turkana and is with Kenya and the other is on the northern end of the coastline with Egypt. In both cases, the administrative boundary does not accord with the political boundary. As a result of political ineptitude since independence, there has been virtually no external support for Sudan. There have also been no influential individuals except for President Nimeri who finally set the civil war on course for current chaos. As a result of decades of conflict and long-term mismanagement, Sudan is ranked first on the Failed State Index.

SYRIA

Location: East Mediterranean coast, at the northern end of Levant coastline. (Map 50)

Area: 185,180 km²   **% Middle East:** 1.1
                    Comparison with Bahrain: 278.5×

Boundaries: **Land: length:** 2,253 km
                    Contiguous countries: Israel 76 km, Jordan 375 km, Lebanon 375 km,
                    Iraq 605 km, Turkey 822 km
                    Maritime: **Coastline:** 193 km
                    Claims: territorial sea 12 nml
                    contiguous zone 24 nml
                    Settlements: Land: Iraq 1916, 1920, 1932 Israel 1916, 1920, 1949
                    Jordan 1916, 1920, 1931 Lebanon 1920, 1945
                    Turkey 1920, 1921, 1929, 1938

Boundary Vulnerability Index: 9

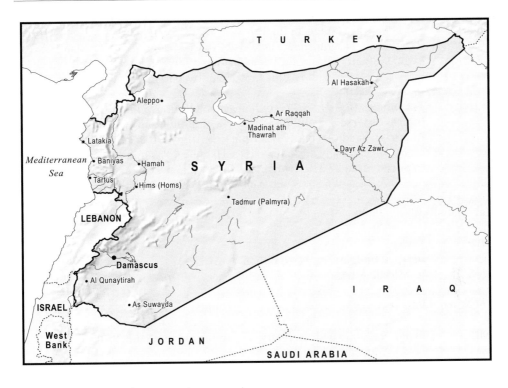

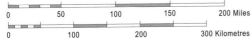

Map 50 Syria

Geography: **Topography:**   Coastal hills and inland the valleys of the rivers Orontes
                                     and Euphrates and desert plains and plateaux.

                  Climate: Mediterranean coastal plain with hot, arid desert inland.

Annual water withdrawal (domestic/industrial/agricultural): 19.95 cu km/yr
              (3%/2%/95%)   **per capita:** 1,048 cu m/yr (2000)

Population: 19.3 million
                  % Middle East: 3.8   **Comparison with Bahrain:** 27.6×
                  Literacy: 79.6%
                  Ethnic groups: Arab 90.3% Kurds, Armenians

Economy: **GDP:** $83.0 billion   **% Middle East:** 2.0
                  Comparison with Bahrain: 3.4× **GDP per capita:** $4,300
                  Key resources: petroleum, agricultural
                  Cropped land: 29.27%   **Irrigated land:** 13,330 km²

Human Development Index Rank: 108

Culture:	Languages:	Arabic (official), Kurdish, Armenian, Aramaic, Circassian, French, English
	Religion:	Muslim (Sunni) 74% other Muslim 16% Christian 10%
	Legal system:	Based on a combination of French and Ottoman civil law with Islamic law in family courts.

Connectivity: **Telephones (fixed):** 3.243 million **Mobiles:** 4.675 million
Internet users: 1.5 million   **% of population:** 7.8
Imports: Saudi Arabia 12.3% China 7.9% Egypt 6.2% UAE 6.0%
        Germany 4.9% Italy 4.9% Ukraine 4.8% Iran 4.5%
Exports: Iraq 27.3% Germany 12.1% Lebanon 9.5% Italy 6.6%
        Egypt 5.3% Saudi Arabia 4.8%

Index of Globalisation: 39.09   **Rank:** 109

Military:	Active:	292,600
	Defence budget:	$1.46 billion
	% Middle East:	1.5

Index of Martial Potency: 6.03   **Rank:** 33

Political: **Type:** Republic under military-   **Independence:** 17 April 1946
        dominant regime
Capital: Damascus
Key individuals: President Hafiz al-Assad

Freedom in the World Rating: **PR:** 7   **CL:** 7   **Status:** NF

Failed State Index: 88.6   **Rank:** 40

Recent events

1958   United with Egypt to form the United Arab Republic.
1961   Separated from Egypt with the re-establishment of the Syrian Arab Republic.
1964   1,000-strong UN Disengagement Observer Force (UNDOF) on Golan.
1967   War with Israel resulted in the loss of the Golan Heights to Israel.
1970   Coup by Hafiz al-Assad.
1976   Troops stationed in Lebanon in what was ostensibly a peacekeeping role.
1981   Golan section of its boundary formally annexed by Israel.
1991   Lost support of the Soviet Union following its dissolution.
2000   Bashar al-Assad followed as President after the death of his father.
2005   Withdrawal of peacekeeping troops from Lebanon.

Key issues

The Golan Heights.
The Kurds.

Hydropolitics of the Tigris–Euphrates and Orontes basins.
International boundaries.
Refugees.
People trafficking.
Drug smuggling.
Terrorism.

Status

Syria is located on the Levant coastline of the eastern Mediterranean between Israel and Turkey. The other contiguous states are Lebanon, Jordan and Iraq. With such neighbours, Syria cannot avoid involvement in the majority of Middle Eastern conflicts. Over the recent past, all of its neighbours other than Jordan have been involved in some form of conflict. Syria is a middle-ranking Middle Eastern state in terms of area, population and GDP although it has a small GDP per capita. Apart from the river valleys, it comprises largely desert plains and plateaux with an increasingly hot and arid desert climate eastwards. Resources are agricultural with minor reserves of petroleum. The population, although predominantly Arab, has significant minorities. This variation is shown in terms of languages, the legal system and religions for which 10 per cent are listed as Christian.

The chief point of contention is the remaining occupation by Israel of the Golan Heights. The area has forty-two Israeli settlements and civilian land-use sites with 18,000 Israeli settlers and about 20,000 Arabs, 18,000 of them Druzes. There is a limited UN presence, the UNDOF.

Among the minorities, the Kurds, although representing only a few per cent of the population, present a series of long-term problems which Syria shares with Turkey, Iran and Iraq. These may be exacerbated if the Kurds in northern Iraq achieve some recognisable form of independence, the basis perhaps for a Kurdish state. A further potential source of instability is the continued dominance of the political apparatus of state by the Alawite minority at the expense of the Sunni majority.

Apart from the river valleys, Syria is a state with serious water shortages. The major river, the Euphrates, rises in Turkey, flows through Syria and reaches the Gulf through Iraq. As the centre of the three co-riparians, Syria is intimately involved in the hydropolitics of the river. The Tigris forms a short length of the eastern boundary of Syria and it is of relatively minor significance. The other major river is the Orontes which flows from Syria to its mouth in Turkey. As the upstream state, Syria has some controlling influence over events as Turkey has to a greater degree over the Euphrates. However, the Orontes is closely tied in with the long-running international boundary problem between Syria and Turkey over Hatay. Apart from this, there are several less significant boundary issues with Lebanon.

Transboundary movements have resulted in a range of issues. There are up to 1.2 million refugees from Iraq and nearly 0.5 million Palestinian refugees as well as some 305,000 internally displaced people, largely from the Golan Heights. Syria is also a destination for the trafficking of women from south and south-eastern Asia and Africa for domestic servitude and from eastern Europe and Iraq for sexual exploitation. Syria is listed as Tier Three of the Watch List. Syria is a transit point for a range of drugs bound particularly for Western markets. It is also a money-laundering centre and has long been associated with terrorism, particularly in support of Pan-Arabism. It is ranked forty on the Failed State Index.

TUNISIA

Location: East end of the *Maghrib* and Atlas Mountain chain with a coastline on the Mediterranean. (Map 51)

Area: 163,610 km²   **% Middle East:** 1.0
Comparison with Bahrain: 246.0×

Boundaries: **Land: length:** 1,424 km
Contiguous countries: Libya 459 km, Algeria 965 km
Maritime: **Coastline:** 1,148 km
Claims: territorial sea 12 nml
EEZ 12nml
contiguous zone 24 nml
Settlements: Land: Algeria 1901, 1970 Libya 1910, 1956, 1970
Maritime: Italy 1971 Libya 1988

Boundary Vulnerability Index: 5

Geography: **Topography:**   Atlas Mountains in the north, plateau in the centre and desert inland in the south.
Climate: Mediterranean modified by altitude in the north; hot, arid desert in the south.

Annual water withdrawal (domestic/industrial/agricultural): 2.64 cu km/yr
(14%/4%/82%)   **per capita:** 261 cu m/yr (2000)

Population: 10.3 million
% Middle East: 2.0   **Comparison with Bahrain:** 14.7×
Literacy: 74.3 %
Ethnic groups: Arab 98%

Economy: **GDP:** $77.16 billion   **% Middle East:** 1.8
Comparison with Bahrain: 3.1×   **GDP per capita:** $7,500
Key resources: –
Cropped land: 30.13%   **Irrigated land:** 3,940 km²

Human Development Index Rank: 91

Culture: **Languages:**   Arabic (official), French
Religion: Muslim 98%
Legal system: Based upon French civil law and Islamic law.

Connectivity: **Telephones (fixed):** 1.268 million **Mobiles:** 7.339 million
Internet users: 1.295 million   **% of population:** 12.6
Imports: France 25.0% Italy 21.9% Germany 9.7% Spain 4.9%
Exports: France 28.9% Italy 20.4% Germany 8.6% Spain 6.1% Libya 4.9%
USA 4.0%

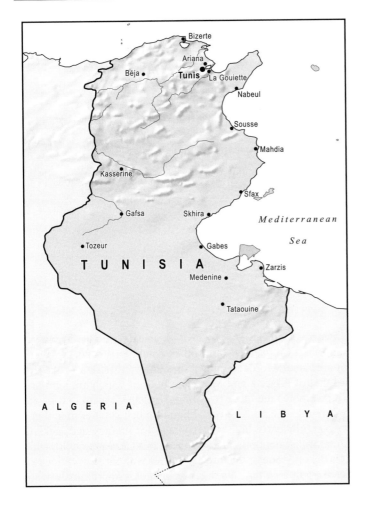

Map 51 Tunisia

Index of Globalisation: 51.81 **Rank:** 73

Military:	**Active:**	35,800
	Defence budget:	$500 million FMA (USA) 8.5 million (0.2% of total)
	% Middle East:	0.5

Index of Martial Potency: 4.13 **Rank:** 87

Political: **Type:** Republic **Independence:** 20 March 1956
Capital: Tunis
Key individuals: President Habibb Bourguibi

Freedom in the World Rating: **PR:** 6   **CL:** 5   **Status:** NF

Failed State Index: 65.6   **Rank:** 122

Recent events

1956   First President, Habibb Bourguibi; held office until 1987 when President Benali took over.
1974   Agreements signed for what was an abortive union with Libya.
1982   Agreement with Libya on the offshore boundary.
1988   Joint exploitation with Libya of the '7 November' oilfield in the Gulf of Gabes.

Key issues

Suppression of Islamist movements.
Trafficking and illegal immigration.

Status

Tunisia is the easternmost of the three states of the *Maghrib* with a coastline on the Mediterranean and a land boundary abutting on to Libya. The only other contiguous state is Algeria. Tunisia is considerably smaller in area than Morocco and of an entirely different order from Algeria. Its population is one-third that of Algeria and Morocco, both of which are similar in population size. Topographically it resembles the other two *Maghrib* states except that its area of hot desert is considerably more limited. The population, although much smaller than that of the other main *Maghrib* states has a significantly higher percentage for literacy and lacks the Berber element in its ethnicity. A major distinction is that Tunisia has a GDP per capita which is almost twice that of Morocco. It is the only one of the three countries without any obvious key resources and it has a more restricted trading pattern. Its defence budget is approximately one-seventh that of Algeria and one-fifth that of Morocco.

Tunisia has virtually no disputes or conflicts internally or externally. Tunisia is only marginally involved in trafficking and the movement of illegal immigrants in sharp contrast to the other two *Maghrib* states. Tunisia has developed a sound economy based quite largely on tourism. As a result, it has the highest living standards in the *Maghrib*.

TURKEY

Location: Bridge between south-eastern Europe and the western Middle East, separated by the Turkish Straits. Coastlines on the Black Sea, the Sea of Marmara and the Mediterranean. (Map 52)

Area: 780,580 km²   **% Middle East:** 4.8
                   Comparison with Bahrain: 1,173.8×

Boundaries: **Land: length:** 2,648 km

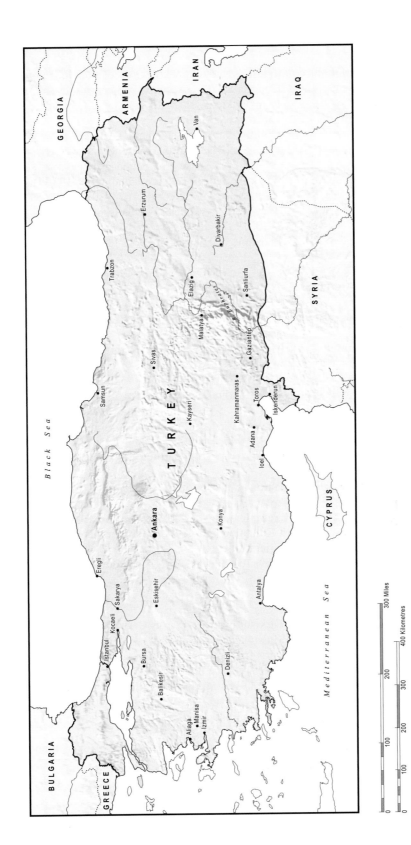

Map 52 Turkey

Contiguous countries: Azerbaijan 9 km, Greece 206 km, Bulgaria 240 km,
 Georgia 252 km, Armenia 268 km, Iraq 352 km,
 Iran 499 km, Syria 822 km

Maritime: **Coastline:** 7,200 km
 Claims: territorial sea 6 nml (Aegean) 12 nml
  (Black Sea and Mediterranean)
  EEZ to agreed boundary (Black Sea)

Settlements: Land: Armenia 1921 Azerbaijan 1921, 1936 Bulgaria
 1879, 1913, 1915 Georgia 1878, 1921 Greece 1829, 1832,
 1880, 1881, 1887, 1920, 1923 Iran 1639, 1847, 1911,
 1913, 1937 Iraq 1925, 1926 Syria 1920, 1921, 1929, 1938
 Maritime: Russia 1973, 1978, 1987 Ukraine 1978, 1986,
 1987, 1993

Boundary Vulnerability Index: 7.5

Geography: **Topography:**   An east-west promontory of mountains and plateaux with
 narrow coastal plains.
 Climate: Mediterranean modified by altitude and continental
 influences.

Annual water withdrawal (domestic/industrial/agricultural): 39.78 cu km/yr
 (15%/11%/74%)   **per capita:** 544 cu m/yr (2000)

Population: 72.1 million
 % Middle East: 14.1   **Comparison with Bahrain:** 101.7×
 Literacy: 87.4%
 Ethnic groups: Turkish 80% Kurdish 20%

Economy: **GDP:** $667.7 billion   **% Middle East:** 16.0
 Comparison with Bahrain: 27.1× **GDP per capita:** $9,400
 Key resources: agricultural
 Cropped land: 33.2%   **Irrigated land:** 52,150 km²

Human Development Index Rank: 84

Culture: **Languages:**   Turkish (official), Kurdish, Dimli, Azeri, Kabardian
 Religion: Muslim (mostly Sunni) 99.8%
 Legal system: Derived from various European, continental legal systems.

Connectivity: **Telephones (fixed):** 18.978 million   **Mobiles:** 52.663 million
 Internet users: 12.284 million   **% of population:** 17.3
 Imports: Russia 12.8% Germany 10.6% China 6.9% Italy 6.2%
  France 5.2% USA 4.5% Iran 4.0%
 Exports: Germany 11.3% UK 8.0% Italy 7.9% USA 6.0% France 5.4%
  Spain 4.4%

Index of Globalisation: 63.45   **Rank:** 44

Military:	Active:	510,600
	Defence budget:	$10.88 billion FMA (USA) 15.0 million (0.4% of total)
	% Middle East:	11.2

Index of Martial Potency: 6.77   **Rank:** 16

Political: **Type:** Republican Parliamentary   **Independence:** 29 October 1923
              Democracy
           Capital: Ankara
           Key individuals: President Turgat Ozal

Freedom in the World Rating: **PR:** 3   **CL:** 3   **Status:** PF

Failed State Index: 74.9   **Rank:** 92

Recent events

1960, 1971, 1980   Military coups.
1974   Invaded Cyprus and helped establish the Turkish Republic of Northern Cyprus.
1984   A separatist insurgency by the PKK.
1991   Gulf War: Turkey cut off the pipeline to Dortyol, severely restricting Iraqi oil exports.
1997   The military organised a "post-modern" coup to remove the Islamic-oriented government.
1999   PKK leader Abdullah Ocalan captured.
       The PKK insurgents largely withdrew to northern Iraq.
2003   Turkish parliament rejected US request to use Turkish bases for Iraq invasion.
2004   The PKK, now known as the People's Congress of Kurdistan or Kongra-Gel (KGK), announced the end of its ceasefire and further attacks in Turkey ensued: this has brought "hot pursuit" incursions by Turkey into northern Iraq, the latest being in 2008.
2006   The Baku-Tblisi-Ceyhan pipeline inaugurated.

Key issues

Strategic location as a key member of NATO, the only one with a shared boundary with the then Soviet Union.
The Turkish Straits, their control, traffic and the transit of tankers.
The Aegean Sea maritime, air and land boundary dispute with Greece.
Cyprus.
The hydrology of the Tigris–Euphrates Basin.
The Kurds, an issue shared with Iran, Iraq and Syria.
Refugees.
Drugs.

Status

Turkey is a relatively large state occupying the whole of the north-west of the Middle East. In a sense, it balances Iran which is the main state of the north-east. Both are remnants of former empires and Islamic but neither is Arab. A major difference is that Iran is a theocracy whereas Turkey, as rebuilt by Ataturk, is aggressively secular. With some 5 per cent of its territory in Europe, Turkey provides a bridge between the Middle East and Europe. As the only Middle Eastern state which is part European and also as a long-serving member of NATO, Turkey has closer ties to Europe than any other Middle Eastern state. However, Cyprus, despite its divided state and close links to Turkey, has acceded to the EU whereas Turkey is in an early stage of negotiation.

Turkey has perhaps the most strategic location in the whole of the Middle East, further enhanced by its control of the Turkish Straits and thereby Black Sea traffic. As exploitation of the Caspian basin oil proceeds, its location will be critical in the debate over pipeline routing. Turkey also controls the headwaters of both the Tigris and the Euphrates and, with Lebanon and parts of Iran, is the only area in the Middle East with adequate water supplies. Indeed, it has sufficient surplus to propose a Peace Pipeline to the Gulf and the Red Sea and to implement Operation Medusa (the large-scale transport of fresh water by sea) to the Turkish Republic of Northern Cyprus and Israel. The importance of Turkey's location was further illustrated during the Gulf War when it was able to cut off Iraq's oil exports and in addition provide the key NATO air bases for Operation Desert Storm. However, during the build up to Operation Iraqi Freedom in early 2003, the Turkish parliament defied the USA and its own armed forces by refusing to endorse the use of Turkish military bases.

Relations with Greece remain fraught as a result of history but obviously in the dispute over the division of Cyprus and the long-standing problem of the Aegean maritime, air and territorial boundaries. The boundary with Armenia remains closed over the issue of Nagorno-Karabakh, a large Armenian exclave in Azerbaijan. Although Iran, Iraq and Syria are involved in the Kurdish issue, Turkey with the vast majority of Kurds within its boundaries, remains the most interested party.

In area, Turkey is the eighth largest state in the Middle East. Its population of 72.1 million places it significantly ahead of Iran but behind Egypt. Its GDP is the second largest in the Middle East, 1.5 times that of Egypt, and almost 1.2 times that of Saudi Arabia. It is exceeded only by that of Iran which is almost 1.3 times larger. Thus, with Egypt and Iran, Turkey is one of the three super-states of the region. Its GDP per capita is more modest and is exceeded by those of Bahrain, Israel and Cyprus, together with those of all the major oil producers in the region except Algeria and Iraq.

The population is 20 per cent Kurdish and almost wholly Muslim but with a variety of languages. As a result of internal conflict with the PKK (KGK), there are as many as 1.2 million internally displaced people in Turkey.

Apart from Israel, Turkey is the most powerful state militarily in the Middle East with an Index of Martial Potency of 6.77 and a rank of 16. Saudi Arabia, with a defence budget three times as large has an Index of Martial Potency of 6.76 and a rank of 17. Israel has a defence budget which is slightly smaller but is greatly enhanced by the addition of US military assistance. It has an index martial potency of 7.67 and a rank of 8.

Turkey is a focal point on the transit route for heroin from south-western Asia to Western Europe and the USA. There are major trafficking organisations in Istanbul and laboratories in remote locations around the city. There is also money laundering.

Turkey is already one of the three super-states of the Middle East, the most militarily powerful Muslim state and the most strategically located of all the states. As links are developed with Israel, the most militarily powerful state in the region, and the republics of central Asia, the position of Turkey can only be enhanced. Issues that threaten Turkey's stability in the future include continuing controversy over the role of Islam in the state, the position of the Turkish Armed Forces relative to the elected institutions of state, and the Kurdish question.

UNITED ARAB EMIRATES (UAE)

Location: A promontory separating the Gulf from the Gulf of Oman. (Map 53)

Area: 83,600 km²   **% Middle East:** 0.5
Comparison with Bahrain: 125.7×

Boundaries: **Land: length:** 867 km
Contiguous countries: Oman 410 km, Saudi Arabia 457 km
Maritime: **Coastline:** 1,318 km
Claims: territorial sea 12 nml EEZ 200 nml
contiguous zone 24 nml
continental shelf to 200 nml or continental
margin
Settlements: Land: none finalised, but reported settlement with Oman
in 2003
Maritime: Iran 1974 Qatar 1969

Boundary Vulnerability Index: 2

Geography: **Topography:**   Mountain belt in north and east, hot desert in the west and
south.
Climate: Hot, arid desert modified by altitude in places.

Annual water withdrawal (domestic/industrial/agricultural): 2.3 cu km/yr
(23%/9%/68%)   **per capita:** 511 cu m/yr (2000)

Population: 4.4 million
% Middle East: 0.9   **Comparison with Bahrain:** 6.3×
Literacy: 77.9%
Ethnic groups: Emirati 19% Other Arab and Iranian 23% South Asian 50%

Economy: **GDP:** $145.8 billion   **% Middle East:** 3.5
Comparison with Bahrain: 5.9×   **GDP per capita:** $55,200
Key resources: petroleum, natural gas
Cropped land: 3.04%   **Irrigated land:** 760 km²

Human Development Index Rank: 39

Culture: **Languages:**   Arabic (official), Persian, English, Hindi, Urdu
Religion: Muslim 96% (Shi'a 16%)
Legal system: Based on a dual system of *Shari'a* and civil courts

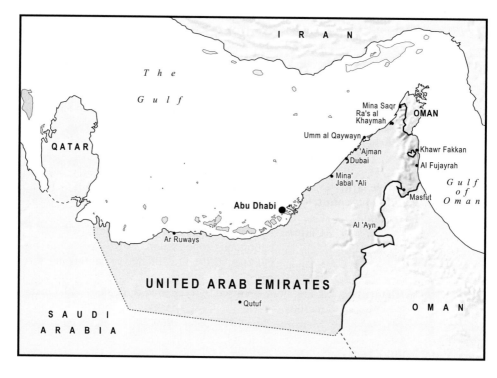

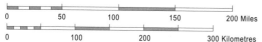

Map 53 United Arab Emirates

Connectivity: **Telephones (fixed):** 1.31 million   **Mobiles:** 5.519 million
Internet users: 1.709 million   **% of population:** 39.0
Imports: USA 11.5% China 11.0% India 9.8% Germany 6.4% Japan 5.8%
UK 5.5% France 4.1% Italy 4.0%
Exports: Japan 25.8% South Korea 9.6% Thailand 5.9% India 4.5%

Index of Globalisation: 70.39   **Rank:** 30

Military: **Active:** 51,000
Defence budget: $10.08 billion
% Middle East: 10.4

Index of Martial Potency: 5.32   **Rank:** 56

Political: **Type:** Federation: specific powers to federal government and to emirates.
Independence: 2 December 1971
Capital: Abu Dhabi
Key individuals: President Zayid bin Sultan Al Nuhayyan

Freedom in the World Rating: PR: 6   **CL:** 6   **Status:** NF

Failed State Index: 51.6   **Rank:** 138

Recent events

1971   The six emirates, Abu Dhabi, Ajman, Fujayrah, Sharjah, Dubai and Umm al Qaywayn merged to form the United Arab Emirates.
1972   The emirate Ra's al Khaymah joined the UAE.
1990   Joined the Coalition in the Gulf War.
1992   The issue of Abu Musa and the Tunbs Islands, all disputed with Iran, resurfaced.
2003   Agreement reportedly signed and ratified with Oman for the entire international boundary including Musandam Peninsula and Al Mahdah enclaves.
2004   Death of the UAE's founding father and first president, Zayid bin Sultan Al Nuhayyan to be succeeded by Khalifa bin Zayid al Nuhayyan, the Ruler of Abu Dhabi.
2008   Agreement with the USA to develop nuclear power for peaceful purposes, the first such agreement with a GCC state.
        Agreement with Saudi Arabia on customs issues in relation to their shared land boundary, the complete delimitation of which has still to be finally agreed.

Key issues

Size of the expatriate workforce: under 20 per cent of the population is indigenous; 73.9 per cent of the population in the 15–64 age group is non-national.
Economic and geographical imbalance between Abu Dhabi and the remaining emirates.
Development of further free trade zones, following the success of Jebel Ali.
The Strait of Hormuz and oil security.
Abu Musa and the Tunbs Islands.
People trafficking.
Drugs.

Status

The United Arab Emirates is located in the centre of the Gulf region, a promontory separating the Gulf from the Gulf of Oman. However, it does not control the Strait of Hormuz as ownership of the nearby Tunbs Islands and Abu Musa is disputed with Iran and the coastline around the Strait itself is an exclave of Oman. The maritime boundary with Qatar was settled but the acceptance of Saudi Arabian corridor between Qatar and the UAE has made that boundary redundant and produced a requirement for new maritime boundaries. Details of the final land boundary with Oman have yet to be published but a point of interest already apparent is that a sizeable fence has been constructed along the boundary in the area of the Buraimi Oasis.

The topography comprises a mountain belt but is predominantly hot desert and water is a key issue necessitating dependence upon desalination. The population of 4.4 million is 19 per cent Emirati. Other Arab nationals and Iranians make up 23 per cent and south Asians 50 per cent. This ethnic make-up is reflected in the languages which include Arabic, Persian, English, Hindi and Urdu. The population is 96 per cent Muslim, 16 per cent of which is Shi'a. The foundation of the state is petroleum and natural gas found in three of the emirates but principally Abu Dhabi. As a result of this and the small population, the GDP per capita is the third largest in the Middle East, behind only those of Qatar and Kuwait and almost twice that of Israel. There is a similarly expansive defence budget which at over $10 billion is marginally less than that of Turkey and seriously exceeded only by that of Saudi Arabia. However, the UAE has an Index of Martial Potency of 5.32 and ranks only fifty-sixth in the world. Although with approximately the same size of defence budget, Turkey has ten times the number of active service personnel. Per head of the military, the UAE defence budget is by some way the largest in the Middle East.

The UAE is a key destination for people trafficked from south and east Asia, eastern Europe, Africa and the Middle East for involuntary servitude and sexual exploitation. An estimated 10,000 women from sub-Saharan Africa, Eastern Europe, south and east Asia, Iraq, Iran and Morocco may be victims of sex trafficking in the UAE. In 2005 a law was passed banning the trafficking of children to be camel jockeys. The UAE is on the Tier Two Watch List. The UAE is an important trans-shipment point for drugs and, given its status as a major finance centre, is vulnerable to money laundering.

WESTERN SAHARA

Location: North-west corner of Africa between Morocco and Mauritania. Its legal status in unresolved. (Map 54)

Area: 266,000 km² **% Middle East:** 1.6
Comparison with Bahrain: 400.0×

Boundaries: **Land: length:** 2,046 km
Contiguous countries: Algeria 42 km, Morocco 443 km,
Mauritania 1,561 km
Maritime: **Coastline:** 1,110 km
Settlements: Land: Algeria 1904 demarcation 1956–8 Mauritania 1900,
1912, 1956 Morocco 1904, 1912
Maritime: dependent upon the resolution of the sovereignty
issue

Boundary Vulnerability Index: 12.5

Geography: **Topography:** Coastal desert rising to uplands in the interior.
Climate: Hot, arid desert.
Annual water withdrawal (domestic/industrial/agricultural): –
Population: 382,600
% Middle East: 0.1 **Comparison with Bahrain:** 54%
Literacy: –
Ethnic groups: Arab, Berber

Economy: **GDP:** – **% Middle East:** –
Comparison with Bahrain: – **GDP per capita:** –
Key resources: phosphates
Cropped land: 0.02% **Irrigated land:** –

Human Development Index Rank: –

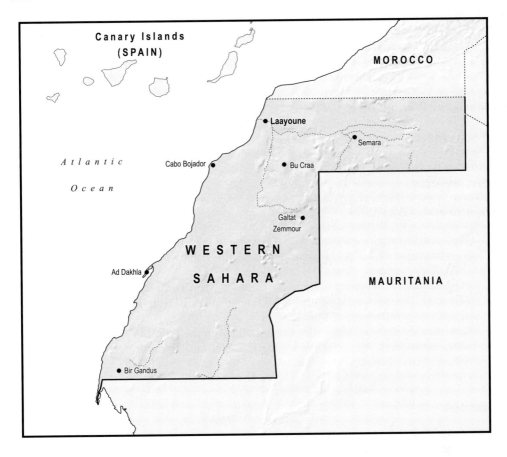

Map 54 Western Sahara

Culture: **Languages:**   Hassaniya Arabic, Moroccan Arabic
                Religion: Muslim
                Legal system: –

Connectivity: **Telephones (fixed):** 2,000   **Mobiles:** –
                Internet users: –   **% of population:** –
                Imports: Included in Morocco
                Exports: Included in Morocco

Index of Globalisation: –   **Rank:** –

Military: **Active:**   –
                Defence budget: –
                % Middle East: –

Index of Martial Potency: –   **Rank:** –

Political: **Type:** Legal status unresolved.   **Independence:** –
                     Capital: –
                     Key individuals: –

Freedom in the World Rating: PR: –   **CL:** –   **Status:** –

Failed State Index: –   **Rank:** –

Recent events

1973   Polisario Front formed.
1976   Two-thirds of Western Sahara (then Spanish Sahara) under virtual annexation by
        Morocco; territory partitioned with the southern third under the control of Mauritania;
        a government in exile, the government of the Sahrawi Arab Democratic Republic
        (SADR), a member of the OAU, established in Algeria.
1979   Withdrawal of Mauritania from its third of Spanish Sahara and replacement by
        Morocco; this was followed by a protracted guerrilla war fought by the Polisario
        against the government of Morocco.
1991   Ceasefire through the UN ended the fighting with the Polisario Front; referendum on
        status postponed. Administered by the UN Mission for the Referendum in Western
        Sahara (MINURSO).

Key issues

Legal status and sovereignty unresolved.
Relationship with Morocco.
Refugees.

Status

Western Sahara is located on the Atlantic coast of North Africa between Morocco and Mauritania. It is not an officially recognised state and remains administered by Morocco. All proposals
put forward by MINURSO have failed. Several states have developed diplomatic relations with
the Sahrawi Arab Democratic Republic represented by the Polisario in exile in Algeria. Other
states recognise Moroccan sovereignty.

Western Sahara is a medium-sized territory with the lowest population of any state or territory in the Middle East. Apart from the coastal fringe, it comprises hot desert and only 0.02 per
cent is cropped. The key resource is phosphates but data on the GDP are not available. There are
102,000 Sahrawi refugees, mostly in camps in the Tindouf area of Algeria. How effectively
Western Sahara could survive as an independent state must be open to question. At present, it is
a dominant factor in the relations between Algeria and Morocco.

YEMEN

Location: South-west corner of the Arabian Peninsula with coastlines on the Red Sea, the Gulf of Aden and the Arabian Sea. (Map 55)

Area: 527,970 km²   **% Middle East:** 3.2
              Comparison with Bahrain: 793.9×

Boundaries: **Land: length:** 1,746 km
                  Contiguous countries: Oman 288 km, Saudi Arabia 1,458 km
                  Maritime: **Coastline:** 1,906 km
                                    Claims: territorial sea 12 nml EEZ 200 nml
                                                continental shelf to 200 nml or continental
                                                margin
                                                contiguous zone 24 nml (Red Sea)
                  Settlements: Land: Oman 1992 Saudi Arabia 1914, 1995, 2000
                                    Maritime: Eritrea 1999 Saudi Arabia 2000

Boundary Vulnerability Index: 2

Geography: **Topography:**   Narrow coastal plain backed by mountains with a hot
                                    desert interior to the north and east.
                  Climate: Hot and arid desert, modified by altitude and maritime
                                    influences. Affected by the Indian monsoon.

Annual water withdrawal (domestic/industrial/agricultural): 6.63 cu km/yr
              (4%/1%/95%)   **per capita:** 316 cu m/yr (2000)

Population: 22.2 million
                  % Middle East: 4.4   **Comparison with Bahrain:** 31.7×
                  Literacy: 50.2%
                  Ethnic groups: Predominantly Arab, some Afro-Arab, South Asians and
                                    Europeans

Economy: **GDP:** $52.61 billion   **% Middle East:** 1.3
                  Comparison with Bahrain: 2.1× **GDP per capita:** $2,400
                  Key resources: petroleum, coffee
                  Cropped land: 3.16%   **Irrigated land:** 5,500 km²

Human Development Index Rank: –

Culture: **Languages:**   Arabic
                  Religion: Muslim including Shaf'i (Sunni) and Zaydi (Shi'a)
                  Legal system: Based upon Islamic law, Turkish law, English common
                                    law and local tribal customary law

Connectivity: **Telephones (fixed):** 968,000   **Mobiles:** 2.0 million
                  Internet users: 270,000   **% of population:** 1.2

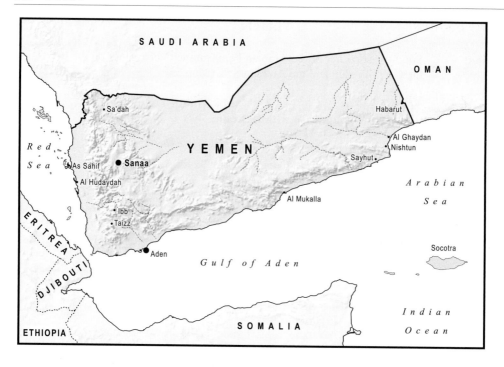

Map 55 Yemen

Imports: UAE 16.4% China 12.8% Saudi Arabia 7.7% Kuwait 5.8%
Brazil 4.5% Malaysia 4.2% USA 4.0%
Exports: China 31.4% India 17.4% Thailand 16.7% South Korea 7.0%
USA 6.7% UAE 4.1%

Index of Globalisation: –   **Rank:** –

Military: **Active:**   66,700
                  Defence budget: $908 million FMA (USA) 8.5 million (0.2% of total)
                  % Middle East: 0.9

Index of Martial Potency: 4.66   **Rank:** 77

Political: **Type:** Republic   **Independence:** 22 May 1990
                  Capital: Sanaa   (Unification Day)
                                                  (North Yemen November 1918,
                                                  South Yemen 30 November 1967)
                  Key individuals: President Ali Salih

Freedom in the World Rating: **PR:** 5   **CL:** 5   **Status:** PF

Failed State Index: 93.2   **Rank:** 24

Recent events

1990   The formal unification of North and South Yemen to produce the Republic of Yemen; North Yemen had been independent since 1918 and Aden, the core of South Yemen, had been a UK Protectorate until 1967.

1991   Gulf War; as a result of Yemeni support for Saddam Hussein, one million workers were expelled from Saudi Arabia.

1992   Petroleum was discovered.

1994   A successionist movement began in the south but was quickly subdued.

1996   Maritime boundary with Eritrea was agreed.

2000   The boundaries with Saudi Arabia were agreed.

Key issues

The Strait of Bab al Mandab.
Relationship with Saudi Arabia.
Petroleum.
Refugees.

Status

Yemen is located in the south-western part of the Arabian Peninsula with coastlines on the Red Sea, the Gulf of Aden and the Arabian Sea. It is the area of *Arabia Felix*, the earliest inhabited part of the Peninsula. Apart from climate, the major advantages of the location are control over the Strait of Bab al Mandab and the possession of a natural, large, deep-water harbour at Aden, adjacent to the strait. The coastal plain, particularly that of the Red Sea, is famous for agriculture, especially coffee which is particularly favoured by the climate. One peculiarity of the agriculture is the concentration on qat which provides a commonly consumed chewable stimulant. The interior is largely hot desert. It is one of the only areas away from the major rivers in the Middle East which has some permanent surface flow.

Yemen is at the upper end of the medium-sized states of the Middle East, being significantly larger than Iraq and Morocco but smaller by roughly the same amount than Afghanistan. The population of 22.2 million places it tenth in size among Middle Eastern countries, larger than Syria but smaller than Saudi Arabia and Iraq. However, Yemen remains a poor country with, apart from Afghanistan, the lowest literacy rate in the Middle East. Its GDP per capita exceeds only those of Mauritania, Palestine (Gaza Strip and the West Bank) and Afghanistan in the region.

The boundary settlement with Saudi Arabia was a major event and sections have now been reinforced by a security barrier to prevent illegal cross-boundary activities. There are some 79,000 refugees from Somalia in the country and this number seems likely to increase.

Yemen is one of the poorest countries in the Middle East with a heavy reliance upon remittances. It was disastrous when, as a result of the Gulf War, the workers were expelled from Saudi Arabia. Hopes rest upon the oil production and possibly the development of Aden as a US Navy base.

Rimland regions

The delimitation of the global area known as the Middle East is itself somewhat contentious. Therefore, the identification of contiguous regions, which in some sense react with the Middle East and for convenience can be termed the Rimland, must be to an extent arbitrary. Some groups of countries, such as those of Central Asia or the Horn of Africa, comprise a recognised region; others, such as parts of south-eastern Europe or south-western Asia, do not. Some regions, such as the Sahel, consist of several countries; others, such as the contiguous part of south-western Asia, comprise only one. Only four of the countries included have no contiguous boundary with a Middle Eastern state. Kazakhstan and Kyrgyzstan are included because they have close historical and current ties with the other three states of Central Asia. All the states of the Horn of Africa are separated from the core of the Middle East by the Red Sea and the Gulf of Aden but whereas Eritrea and Ethiopia have a common border with Sudan to the west, Djibouti and Somalia have no contiguous boundaries with Middle Eastern states. The one state not included, which has a contiguous boundary, is China. The boundary stretches for 76 km through the Karakoram Range at the eastern end of the Wakhan Panhandle. This boundary accounts for 0.3 per cent of the total boundary length of China. Given the almost total inaccessibility of this boundary, it is felt that China is effectively no more in the Rimland of the Middle East than Russia or India. The influence of these three great powers is far more pervasive in the Middle East and more significant in the geopolitics than that of any Rimland region. These influences are considered, as appropriate, in Sections C, D and E.

For each Rimland region, the general characteristics are described with a focus upon links to the Middle Eastern states. The states of each Rimland region are then discussed. A table of key points, illustrating the basic strength of the state as an actor upon the world stage, is followed by a description of the links and interactions with the Middle East. These descriptions illustrate how the Middle Eastern influence is attenuated. For example, taking Islam as a characteristic, the effect of the Middle East declines sharply in Bulgaria and especially Greece, whereas, in the case of Pakistan, attenuation only commences beyond its boundary with India. The Middle East can also be approximately delimited by its Rimland seas. These points help illuminate the fact that the Middle East is defined by a set of mental constructs rather than by any geographical boundaries.

SOUTH-EASTERN EUROPE

For convenience, Greece and Bulgaria are considered together as a region (Map 56) but the designation of south-eastern Europe is vague in that such a region also includes European Turkey and several other states. However, only Greece and Bulgaria have common boundaries with a Middle Eastern state, though these two differ in terms of basic geography in that one is a

Map 56 South-eastern Europe

Mediterranean and the other a Black Sea state. Despite Muslim influence to the west, notably in Bosnia, this region is effectively at the edge of the Islamic world and is distinct from the Middle East. Indeed, the dominant issue within the region has been the continuing tension between Greece and Turkey, particularly over the delimitation of the Aegean Sea but also over the future of Cyprus. The island of Cyprus illustrates the problem of identity in the region in that while, by geographical location, it belongs to Asia and the Middle East, as a result of the Greek majority population, it is commonly perceived as European.

Bulgaria

Area:	110,910 km²	**Comparison with Bahrain:**	116.8×
Population:	7.263 million	**Comparison with Bahrain:**	10.2×
GDP $:	86.32 billion	**Comparison with Bahrain:**	3.2×
GDP per capita $:	11,300		
Active Military:	51,000	**Index of Martial Potency:**	5.01
Contiguous boundary:	Turkey 240 km	**Geopolitical Index:**	4.5

Bulgaria is a small country in area, within the Rimland only Djibouti and the states of the Southern Caucasus being smaller. In population, it is middle ranking but it has a GDP which is higher than any of the other Rimland states apart from Pakistan, Greece and Kazakhstan. Its GDP per capita is second only to that of Greece. Within the Rimland, it is of more than average military strength and its Index of Martial Potency is exceeded only by Pakistan, Greece and Kazakhstan.

Bulgaria has no international disputes and is not implicated in people trafficking. However, it is a major trans-shipment point for heroin from south-west Asia and South American cocaine, both drugs destined for the European market. There is some money laundering.

Bulgaria's links with the Middle East are predominantly through Turkey. Of the population 9.4 per cent is ethnically Turkish and 9.6 per cent speaks Turkish. Turkey provides 7.1 per cent of Bulgaria's imports and takes 11.6 per cent of its exports. Some 12.2 per cent of the population is listed as Muslim.

Therefore, Bulgaria has clear Middle Eastern connections but they are limited in scope.

Greece

Area:	131,940 km²	**Comparison with Bahrain:**	198.4×
Population:	10.723 million	**Comparison with Bahrain:**	15.1×
GDP $:	324.6 billion	**Comparison with Bahrain:**	13.2×
GDP per capita $:	29,200		
Active Military:	147,100	**Index of Martial Potency:**	6.37
Contiguous boundary:	Turkey 206 km	**Geopolitical Index:**	12

Although rather larger than Bulgaria, Greece is still a middle-ranking state in area. Its population is also in the medium range but, in the Rimland, the GDP is second only to that of Pakistan. Its GDP per capita is by far the largest in the Rimland, more than 2.5 times that of the second ranking state, Bulgaria. In terms of military strength, it ranks fourth in the Rimland behind Ethiopia, Eritrea and Pakistan which has more than four times the military in terms of numbers. Its Index of Martial Potency is exceeded marginally by that of Pakistan.

Greece has less obvious Middle Eastern connections than Bulgaria, depending upon how its relationship with Cyprus is viewed. Of the population 1.3 per cent is Muslim. Among its export partners, Cyprus takes 5.4 per cent and Turkey 5.1 per cent.

Greece has a range of international disputes, the most complex concerning the maritime, air and territorial boundary delimitation disputes with Turkey in the Aegean Sea. The issue of

Cyprus remains a point of contention between Greece and Turkey but there are signs that attitudes are softening. As a result of history and national pride, Greece continues to dispute the use of the name Macedonia to identify the state of that name. There are continuing refugee problems with Albanians and Greece provides an entry point for traffickers of cannabis, heroin and cocaine into Europe. There is related money laundering.

Greece has very little obvious interaction with the Middle East other than its continuing confrontation with Turkey and its links with Cyprus.

SOUTHERN CAUCASUS

The region comprises three small states which are located in the south of the Caucasus, the land between the Black Sea and the Caspian Sea. Each has at least one common boundary with a Middle Eastern state (Map 57). The region is the key corridor between Central Asia and Europe and is particularly important for the transport of petroleum by pipelines. One of the states, Azerbaijan, is a major Caspian Basin oil state and as the Caspian basin as a whole becomes more important in global petroleum production, so the significance of the region is likely to increase. All three states were part of the former Soviet Union and all belong to the CIS. However, the

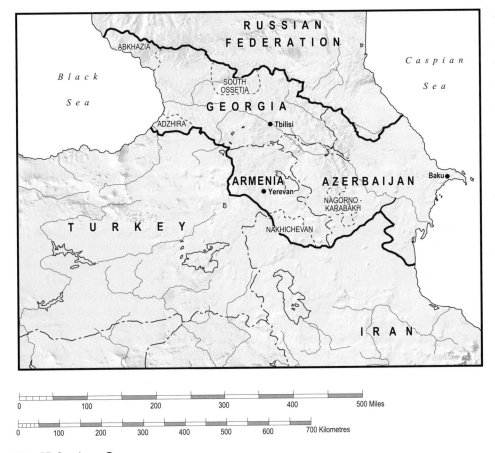

Map 57 Southern Caucasus

three states vary considerably in history and development and the region has been one of continuing conflict. In particular, there has been a long-standing dispute between Armenia and Azerbaijan over the Nagorno-Karabakh, a predominantly Armenian-populated region, entirely within Azerbaijan. The conflicts in the third state, Georgia, have been with Russia and internal resistance to the rule of the central government. While all three have common boundaries with each other, Georgia is a Black Sea state, Azerbaijan is a Caspian Sea state and Armenia is landlocked.

Armenia

Area:	29,800 km²	**Comparison with Bahrain:**	44.8×
Population:	2.972 million	**Comparison with Bahrain:**	4.2×
GDP $:	16.94 billion	**Comparison with Bahrain:**	69%
GDP per capita $:	5,700		
Active Military:	43,641	**Index of Martial Potency:**	3.66
Contiguous boundaries:	Iran 35 km	**Geopolitical Index:**	3
	Turkey 268 km		8

Other than Djibouti, which is in effect a micro-state, Armenia is the smallest state in the Rimland in terms of both area and population. However, its GDP is significantly above the average for the region and its GDP per capita is exceeded by only five states in the Rimland. Statistics for its military strength and Index of Marital Potency place it in the middle rank. Both boundaries with Middle East countries are politically active, particularly the Turkish boundary which was closed by Turkey, following an economic blockade, because of the Armenian occupation of Nagorno-Karabakh and other areas.

With no Muslims and no speakers of any of the major Middle Eastern languages, Armenia has few obvious links with the Middle East. With regard to ethnicity, religion and language, the main connection is through Yezidi, a religion linked ethnically with the Kurds. However, statistics for the Yezidi influence vary only from 1–1.3 per cent. Among the more minor trading partners are Israel and Iran which account respectively for 6.6 per cent and 4.6 per cent of the exports and 4.5 per cent and 5.4 per cent of the imports. Therefore, Middle Eastern influences are very limited in Armenia, considerably less than in Azerbaijan and significantly less than in Georgia.

The key international dispute is with Azerbaijan over the Nagorno-Karabakh. Armenia supports ethnic Armenian secessionists in Nagorno-Karabakh and, since the early 1990s, has occupied militarily 16 per cent of Azerbaijan. The net result has been a relatively large-scale population movement: 800,000 mostly ethnic Azerbaijanis leaving the occupied lands and 230,000 ethnic Armenians moving from Azerbaijan to Armenia. Azerbaijan is still seeking a fixed and safe route to its exclave, Nakhichevan. Armenian groups in the Javakheti area of Georgia have demanded greater autonomy. The number of refugees in the country is estimated at just under 220,000, all from Azerbaijan. The number of internally displaced persons is put at 8,400.

A particularly unsavoury link with the Middle East is the fact that Armenia is a major source, transit route and destination for females trafficked for sexual exploitation largely to the UAE and Turkey. Armenia has failed to show evidence of interest in efforts to curb this trade and is on the Tier Two Watch List. Armenia is a minor transit point for the movement of drugs from south-west Asia to Russia and, to a lesser extent, to the rest of Europe.

Azerbaijan

Area:	86,600 km²	**Comparison with Bahrain:**	130.2×
Population:	8.12 million	**Comparison with Bahrain:**	114.6×
GDP $:	59.71 billion	**Comparison with Bahrain:**	2.4×
GDP per capita $:	7,500		
Active Military:	66,740	**Index of Martial Potency:**	4.21
Contiguous boundaries:	Iran 432 km	**Geopolitical Index:**	8
	Iran (Nakhichevan Exclave)		8
	Turkey 9 km		0

Azerbaijan is the largest state of the Southern Caucasus, almost three times as large as Armenia. Its population is relatively small but is 1.7 times that of Georgia and 2.7 times that of Armenia. In terms of GDP, its statistic is far above that of the other two states of the region and indeed is exceeded only by the states of south-eastern Europe, Pakistan, Kazakhstan and Ethiopia within the Rimland. Its GDP per capita is exceeded within the Rimland only by those of Greece, Bulgaria, Kazakhstan and Turkmenistan. Azerbaijan is militarily relatively strong and, as a result of its petroleum deposits, it is one of the few Rimland states to be of any real global significance. With Azeri populations dominating the borderlands, the boundaries with Iran are relatively volatile.

Despite the lack of ethnic or linguistic links to the Middle East, the population of Azerbaijan is 93.4 per cent Muslim. Its trade pattern is almost exclusively non-Middle Eastern with only 9.9 per cent of imports coming from Turkey and no other country from the Middle East featuring on the trade list.

There is a long-running dispute with Armenia over the ethnic Armenian secessionists who dominate the Azerbaijani enclave of Nagorno-Karabakh. This has resulted in more than 800,000 largely ethnic Azerbaijanis being driven from the military occupied lands and Armenia and about 230,000 ethnic Armenians being driven the other way from their homes in Azerbaijan into Armenia. Azerbaijan also seeks a reliable transit route to its exclave, Nakhichevan. There is continuing mediation in the dispute by the Organisation for Security and Cooperation in Europe (OSCE). There is a dispute with Iran over Azerbaijan's Caspian petroleum exploration in what are considered to be disputed waters. Azerbaijan, Kazakhstan and Russia have ratified Caspian Seabed delimitation treaties based on equidistance but Iran insists on an even one-fifth allocation between the five littoral states. Bilateral talks continue with the other Caspian littoral state, Turkmenistan, over seabed and Caspian oilfield delimitation. There is a continuing relatively minor dispute with Georgia over the alignment of their common boundary at certain crossing points.

The number of refugees is listed at 2,800, all from Russia, and internal displaced persons at 580,000–690,000, resulting from the conflict over Nagorno-Karabakh. With regard to drugs, there is limited cultivation of cannabis and opium mostly for the CIS and Azerbaijan is a transit point for opiates to Russia and, to a lesser extent, to the rest of Europe.

Georgia

Area:	69,700 km²	**Comparison with Bahrain:**	104.8×
Population:	4.646 million	**Comparison with Bahrain:**	6.6×
GDP $:	17.88 billion	**Comparison with Bahrain:**	73%
GDP per capita $:	3,800		
Active Military:	11,320	**Index of Martial Potency:**	3.26
Contiguous boundary:	Turkey 252 km	**Geopolitical Index:**	3

A small state with a low population total, Georgia ranks between the other two states of the Southern Caucasus for both statistics. However, in the context of the Rimland, only Djibouti and Armenia have a smaller area and only the same two states have a smaller population. However, its GDP is larger than that of eight Rimland states and its GDP per capita is greater than that of thirteen other Rimland states. Therefore, like the other states of the Southern Caucasus, the GDP of Georgia is out of phase with its area and population size. Although it has an Index of Martial Potency similar to that of Armenia and greater than that of seven other Rimland states, it has a low military strength, only marginally above that of Djibouti. Within the Rimland, its military strength exceeds only those of Somalia, which has no military, Djibouti, Niger, Mali and Tajikistan. Unlike the other states of the Southern Caucasus its common boundary with the Middle East has a low geopolitical index.

Georgia's ethnic and linguistic links with the Middle East are somewhat tenuous, being through its Azeri minority, which also constitutes a significant minority in Iran. However, 9.9 per cent of the population is Muslim and Turkey is an important trading partner accounting for 16.9 per cent of exports and 12.3 per cent of imports.

Disputes are both internal and external. Government control varies throughout the country and there are two notable breakaway regions, Abkhazia and South Ossetia. These two territories are ruled by *de facto* unrecognised governments supported by Russia and there are Russian-led peacekeeping operations in both regions. Externally, there are disagreements with Russia on the delimitation of about 20 per cent of their common border including a number of particularly volatile areas such as the Pankisi Gorge and the Argun Gorge. These disputes are monitored by the OSCE and there has been a UN observer mission in Georgia since 1993. Armenian groups in the Javakheti region seek greater autonomy and certain adjustments to the boundary with Azerbaijan are the subject of discussions.

Between 220,000 and 240,000 internally displaced persons are listed as a result of migration from Abkhazia and South Ossetia. As in the other states of the region, there is some limited cultivation of drugs and Georgia is a trans-shipment point for opiates from Central Asia to Russia and Western Europe.

CENTRAL ASIA

The states of Central Asia are all former republics of the Soviet Union, all have large numbers of Russian speakers and all are landlocked (Map 58). Although only three of the states have common boundaries with states of the Middle East, the five are included because they constitute a recognised region. Furthermore, as a result of their history, they remain relatively closely integrated. The most detached, Kazakhstan, is linked to the Middle East through its significance as a Caspian basin oil producer and the pipelines, potential and actual, which connect it with the Middle East.

The major link to the Middle East is through Islam and Muslims predominate in all five states.

Kazakhstan

Area:	2,717,300 km²	**Comparison with Bahrain:**	4,086.2×
Population:	15.285 million	**Comparison with Bahrain:**	21.6×
GDP $:	143.1 billion	**Comparison with Bahrain:**	5.8×
GDP per capita $:	9,400		
Active Military:	65,800	**Index of Martial Potency:**	5.38
Contiguous Boundaries:	–	**Geopolitical Index:**	–

Map 58 Central Asia

Kazakhstan is by far the largest Rimland state being more than twice the size of Chad, the second largest. It is indeed the largest landlocked state in the world. For its area, its population is low and, within the Rimland, is exceeded by three states. Nonetheless, its GDP is easily the highest among the Central Asian states, more than 2.5 times the second largest GDP, that of Uzbekistan. Within the Rimland as a whole, its GDP is exceeded only by that of Pakistan, a state with almost eleven times the population. Kazakhstan's GDP per capita is behind only those of Greece and Bulgaria. Although its military strength is modest, its Martial Potency is exceeded, among Rimland states, only by those of Pakistan and Greece. In terms of military strength, Kazakhstan is well behind Pakistan, Eritrea, Ethiopia and Greece and marginally behind Azerbaijan. The military strength of Pakistan is 9.4 times that of Kazakhstan.

Kazakhstan has no geographical link with the Middle East but linguistically, Kazakh is a member of the Turkic family of languages. The most obvious connection is however through the fact that the predominant religion, practised by 47 per cent of the population, is Islam. Kazakhstan's trade pattern includes no Middle Eastern partners.

With regard to international disputes, all concern boundaries. The boundary delimitation with Kyrgyzstan (2001) has yet to be ratified. Boundary demarcation is underway with Turkmenistan, Uzbekistan and Russia and was completed with China in 2002. With regard to the seabed, the boundary with Turkmenistan in the Caspian Sea is under negotiation while those with Azerbaijan and Russia have been ratified as equidistant lines. The division of the water column has yet to be decided.

The number of refugees, all from Russia, is listed as 5,000. There is significant illegal drug cultivation, especially of cannabis, particularly for the CIS markets and Kazakhstan is a transit point for narcotics bound for Russia and Western Europe from south-west Asia.

Kyrgyzstan

Area:	198,500 km²	**Comparison with Bahrain:**	298.5×
Population:	5.284 million	**Comparison with Bahrain:**	7.5×
GDP $:	10.73 billion	**Comparison with Bahrain:**	44%
GDP per capita $:	2,100		
Active Military:	12,500	**Index of Martial Potency:**	3.97
Contiguous boundaries:	–	**Geopolitical Index:**	–

Kyrgyzstan is a medium-sized country, in the median position among the states of the Rimland. Its population is marginally larger than that of Turkmenistan and these two countries have the lowest population in Central Asia, under a third that of Kazakhstan and a fifth that of Uzbekistan. Among the Rimland states, only Tajikistan, Niger, Somalia, Eritrea and Djibouti have a lower GDP. Although very low, at $2,100, the GDP per capita is higher than that of Tajikistan and Uzbekistan in Central Asia and is higher than all the states of the Sahel and the Horn of Africa. Military strength is very low, exceeding only those of Djibouti, Tajikistan, Mali, Niger and Somalia. For its size, it has a relatively high Martial Potency. Kyrgyzstan is one of the only four states in the Rimland without a common boundary with a Middle Eastern state.

The main link with the Middle East is that 75 per cent of the population is Muslim but also the language is of the Turkic family. Kyrgyzstan has no Middle Eastern import partners but the UAE is responsible for 35.8 per cent of its exports.

Kyrgyzstan's main international disputes concern boundaries. The boundary delimitation with Kazakhstan in 2001 has yet to be ratified while the completion of boundary delimitation with Tajikistan is delayed by disputes in the Isfara Valley. There remains some 130 km of boundary with Uzbekistan which is as yet undelimited as a result of disputes about enclaves. Like Kazakhstan, Kyrgyzstan has limited illicit cultivation of cannabis and opium for the CIS markets and is a transit point for narcotics from south-west Asia to Russia and the rest of Europe.

Tajikistan

Area:	143,100 km²	**Comparison with Bahrain:**	215.2×
Population:	7.077 million	**Comparison with Bahrain:**	10×
GDP $:	9.521 billion	**Comparison with Bahrain:**	39%
GDP per capita $:	1,300		
Active Military:	7,600	**Index of Martial Potency:**	2.75
Contiguous boundary:	Afghanistan 1,206 km	**Geopolitical Index:**	5

Tajikistan is of medium size within the Rimland but is the smallest of the Central Asian states. Its population of seven million exceeds those of Kyrgyzstan and Turkmenistan but is only just over a quarter that of Uzbekistan. Its GDP is the lowest of the Central Asian states, a little below that of Kyrgyzstan and of a different order of the other three states. Its GDP per capita is the lowest of the Central Asia states and exceeds only those of Niger, Mali and the states of the Horn of Africa within the Rimland. Its military strength is by some distance the lowest in Central Asia and is well below that of Djibouti, Chad and Senegal. The Index of Martial Potency is also the lowest in Central Asia and, in the Rimland, is greater only than those of Chad, Senegal and Djibouti. There is a common boundary with Afghanistan which has a Geopolitical Index of 5, mainly as a result of political turbulence, in that geographical accessibility is very low.

Of the population 90 per cent is Muslim with a ratio of seventeen Sunni to one Shi'a and the language is of the Iranian group. There are further Middle East connections in the trading pattern, although these are relatively minor. Turkey is responsible for 4 per cent of the imports and 12.2 per cent of the exports and Iran for 5.2 per cent of the exports. Tajikistan has been beset by conflict since its independence following the disintegration of the Soviet Union in 1991. The major international disputes concern boundaries and drug trafficking. The boundary with China was delimited in 2002 and agreement was reached to begin demarcation in 2006. Talks continue with Uzbekistan to delimit the boundary and to remove minefields, and disputes in the Isfara Valley have delayed delimitation with Kyrgyzstan.

Tajikistan is a major transit country for narcotics from Afghanistan bound for Russia and to a lesser extent Western Europe. There is limited illicit cultivation of opium. However, Tajikistan seizes approximately 80 per cent of all drugs captured in Central Asia and stands third world-wide in the seizure of opiates.

Turkmenistan

Area:	488,100 km²	**Comparison with Bahrain:**	734×
Population:	5.097 million	**Comparison with Bahrain:**	7.2×
GDP $:	42.84 billion	**Comparison with Bahrain:**	1.7×
GDP per capita $:	8,500		
Active Military:	26,000	**Index of Martial Potency:**	3.65
Contiguous boundaries:	Afghanistan 744 km	**Geopolitical Index:**	6
	Iran 992 km		1.5

Turkmenistan is a very large state, second only in area among the states of Central Asia to Kazakhstan. However, it has a population of only five million, the lowest in Central Asia. Despite its importance within the Caspian Sea basin for oil and natural gas, its GDP is only one-third that of the other main oil state of Central Asia, Kazakhstan. Its GDP per capita is slightly lower than that of Kazakhstan but well ahead of the other three Central Asian states. Turkmenistan's military strength of 26,000 is well ahead of that of Tajikistan and Kyrgyzstan but, within the remainder of the Rimland, is low exceeding only those of Georgia, Djibouti and the states of the Sahel. However, like Kyrgyzstan, the Martial Potency is relatively high for the military strength. Turkmenistan has long boundaries in common with Afghanistan and Iran and these have geopolitical indices respectively of 6 and 1.5.

The key link to the Middle East is the fact that 89 per cent of the population is Muslim and the language is of the Turkic family. There are also well-developed trade links. Iran takes 16.2 per cent of exports while the UAE, Turkey and Iran supply respectively 13.6 per cent, 9.8 per cent and 6.7 per cent of the imports.

The main international disputes are over water and boundaries. Owing to the requirements of the cotton crop, there is a continuing confrontation with Uzbekistan over the allocation of water from the Amu Darya. Boundary demarcation with Kazakhstan commenced in 2005. However, the Caspian Sea delimitation with Kazakhstan, Azerbaijan and Iran remains undecided as a result of Turkmenistan's indecision. Turkmenistan is a transit country for narcotics from Afghanistan bound for Russia and Western Europe.

Uzbekistan

Area:	447,400 km²	**Comparison with Bahrain:**	672.8×
Population:	27.789 million	**Comparison with Bahrain:**	39.2×
GDP $:	55.75 billion	**Comparison with Bahrain:**	2.3×
GDP per capita $:	2,000		
Active Military:	55,000	**Index of Martial Potency:**	4.97
Contiguous boundary:	Afghanistan 137 km	**Geopolitical Index:**	8

Uzbekistan is a large state in area and has by far the largest population in Central Asia. Although only 39 per cent that of Kazakhstan, the GDP of Uzbekistan is exceeded within the remainder of the Rimland only by those of Greece, Pakistan, Ethiopia and Bulgaria. However, at $2,000 the GDP per capita is very low, marginally behind that of Kyrgyzstan. Uzbekistan's military strength places it ahead of Bulgaria and, within the Rimland, behind Pakistan, Eritrea, Ethiopia and Greece. For the Rimland, the Martial Potency of 4.97 is high. Indeed, only the figures for Pakistan, Greece, Kazakhstan and Bulgaria are higher. Uzbekistan has a common boundary with Afghanistan which has seen a good deal of political turbulence and has a Geopolitical Index of 8.

As with the other states of Central Asia, the main link with the Middle East is through religion and language. Of the population 88 per cent is Muslim and the major language belongs to the Turkic group. There is only a minor Middle Eastern element in the trading pattern, with Turkey supplying 4.5 per cent of imports and taking 7.5 per cent of exports.

Uzbekistan has international disputes concerning water, boundaries, refugees, people trafficking and drugs. There is a long-running dispute with Turkmenistan over the allocation of water from the Amu Darya. Demarcation of the boundary with Kazakhstan began in 2004 but the delimitation of the boundary with Kyrgyzstan is characterised by serious disputes about enclaves. There are said to be over 39,000 refugees from Tajikistan within the country and 3,400 internally displaced people. Uzbekistan is fraught with people trafficking issues and appears to have made virtually no attempt to address the situation. No national action plan has been approved. The country is a source and a transit for women trafficked to Asia and the Middle East for sexual exploitation. They come mainly from other Central Asian countries and China. Men are trafficked for forced labour in the Ukraine, Russia, Kazakhstan and Kyrgyzstan. Both men and women are also trafficked within Uzbekistan. As a consequence of its lack of commitment to the eradication of trafficking, Uzbekistan is placed on Tier Three of the Watch List.

There is limited illicit cultivation of cannabis and opium and Uzbekistan is a transit country for narcotics from Afghanistan bound for Russia and, to a lesser extent, Western Europe.

SOUTH-WEST ASIA

As normally defined, south-west Asia includes the Indian subcontinent. However, only Pakistan (Map 59) can be considered to lie within the Middle Eastern Rimland and its own long-running disputes with India preclude much Middle Eastern influence from that country. Pakistan has risen to prominence particularly as a result of the war in Iraq and the growth of international terrorism, most notably through the development of Al Qaeda. Terrorism and transboundary movements between Pakistan and the Middle East are inextricably linked as is the movement of labour into the oil-rich Middle Eastern countries. The other major link with the region is, of course, Islam.

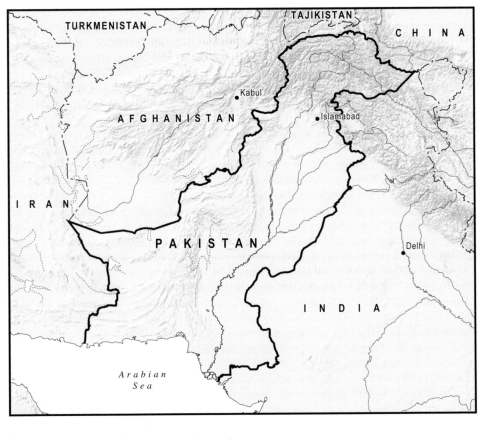

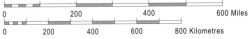

Map 59 Pakistan

Pakistan

Area:	803,970 km²	**Comparison with Bahrain:**	1,298.9×
Population:	164,742 million	**Comparison with Bahrain:**	232.5×
GDP $:	437.5 billion	**Comparison with Bahrain:**	17.8×
GDP per capita $:	2,600		
Active Military:	619,000	**Index of Martial Potency:**	6.69
Contiguous Boundaries:	Afghanistan 2,430 km	**Geopolitical Index:**	7.5
	Iran 909 km		6

In area, Pakistan is exceeded within the Rimland only by the very large, million plus square kilometre states. These comprise Chad, Niger, Mali and Ethiopia. Its population of almost 165 million places it among the more populous states in the world. This population figure is well over twice the size of the second largest Rimland state, Ethiopia, itself considered highly populated by global standards. Its GDP is 1.3 times that of Greece and, the two together are far higher than those of any other Rimland state. However, in the case of Pakistan the GDP per capita, $2,600, is very low, particularly when compared with Greece which has a GDP per capita of $29,200. Indeed, the figure is well below that for Bulgaria, the states of the Southern Caucasus, Kazakhstan and Turkmenistan. Pakistan's military strength is of a different order from that of any other Rimland state. It has more than three times the military strength of the next highest country, Eritrea. As would be expected, the Index of Martial Potency of 6.69 is high by global standards. Only Greece among Rimland states has a figure which exceeds 6.0. Pakistan has two very long boundaries in common with Middle Eastern states. Although both are the scenes of considerable political turbulence, their indices are relatively low owing to their geographical inaccessibility.

The obvious link with the Middle East is through religion: 97 per cent of the population is Muslim (77 per cent Sunni and 23 per cent Shi'a). Pakistan is one of only two Rimland countries to have agreed a maritime boundary with a Middle Eastern state, in this case Oman. Pakistan also interacts significantly with the Middle Eastern states through trade. Important import partners are: Saudi Arabia (10.4 per cent), UAE (9.7 per cent) and Kuwait (4.7 per cent). The UAE takes 9.1 per cent of Pakistan's exports and Afghanistan 7.7 per cent. Pakistan has international boundary disputes with China and India but perhaps the major issue is that of Kashmir. Despite several local conflicts, India and Pakistan have maintained their ceasefire in Kashmir since 2004. There is a long-running dispute with India over the allocation of the water of the River Indus and its tributaries. The two countries also have a maritime boundary dispute at the Rann of Kutch. Pakistan has tried to control movement across the boundary with Afghanistan and portions of the boundary have a fence together with minefields.

There are still over one million Afghan refugees in the country, 2.3 million having been repatriated. There are also 34,000 internally displaced people as a result of government military action and natural disasters. Pakistan is important in global drugs trafficking. Opium cultivation is estimated at 800 hectares and the country is a key transit point for drugs from Afghanistan including heroin, opium, morphine and hashish bound for the Gulf States, Africa and the West. Pakistan is also the scene for financial crimes related to drug trafficking, terrorism, corruption and smuggling. It is ranked twelfth on the Failed State Index.

HORN OF AFRICA

The four states which comprise the Horn of Africa are separated from the Arabian Peninsula by the Gulf of Aden and the Red Sea (Map 60). However, both Eritrea and Ethiopia have a common boundary with Sudan. Ethiopia, by far the largest state in the region, is landlocked. Somalia is the most frequently quoted example of a failed state. Djibouti has strong colonial links and is still dominated by France. Within the region, there is turmoil in the form of various civil wars in Somalia while the boundary conflict between Ethiopia and Eritrea has yet to be resolved.

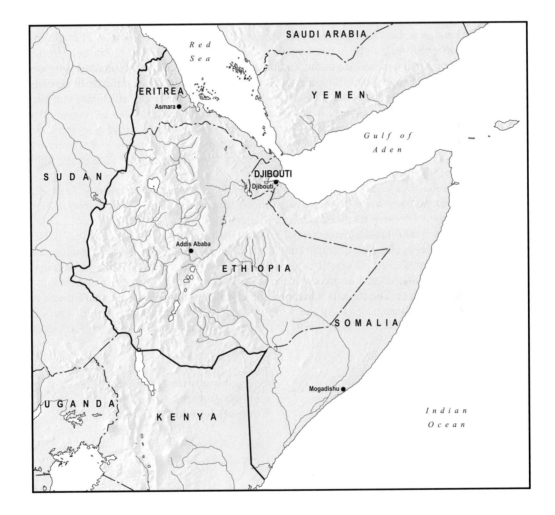

Map 60 Horn of Africa

Djibouti

Area:	23,000 km²	**Comparison with Bahrain:**	34.6×
Population:	0.496 million	**Comparison with Bahrain:**	70%
GDP $:	619 million	**Comparison with Bahrain:**	2.5%
GDP per capita $:	1,000		
Active Military:	10,950	**Index of Martial Potency:**	2.56
Contiguous boundaries:	–	**Geopolitical Index:**	–

In an area of only 23,000 km², Djibouti is by some distance the smallest state in the Rimland. It is about three-quarters the size of Armenia. Despite the fact that it covers some thirty-five times the area of Bahrain, it is, by the standards of the Rimland, effectively a micro-state. The population of under half a million is only 70 per cent that of the population of Bahrain and only 17 per cent of that of the next smallest state in the Rimland, Armenia. It is the only Rimland state the GDP of which does not reach one billion but its GDP per capita of $1,000 comfortably exceeds those for Somalia and Niger and equals the figure for Eritrea and Ethiopia. The military strength of Djibouti is one of the lowest in the Rimland, only Niger, Mali and Tajikistan having smaller totals. As would be expected, the Index of Martial Potency is low, only Niger among all the Rimland states has a lower figure. However, the military is supplemented by the French military and there is a large US military base. Djibouti is considered to be a frontline state in the fight against global terrorism.

Djibouti has close links with the Middle East. Ethnically, part of the population is Arab, 94 per cent is Muslim and Arabic is one of the two official languages. Saudi Arabia supplies 21.4 per cent of imports and Yemen takes 3.4 per cent of exports.

There are almost 9,000 refugees from Somalia in UNHCR camps in Djibouti. However, the main source of conflict is over people trafficking. Djibouti is a source, transit and destination for women and children trafficked for sexual exploitation and possibly forced labour. Indeed, there are close links with Ethiopia and Somalia for trafficking and there is a well-known Ethiopia–Djibouti trucking corridor. The main destinations are Middle Eastern countries and Djibouti is on the Tier Two Watch List.

Eritrea

Area:	121,320 km²	**Comparison with Bahrain:**	182.4×
Population:	4.907 million	**Comparison with Bahrain:**	6.9×
GDP $:	4.471 billion	**Comparison with Bahrain:**	18.2%
GDP per capita $:	1,000		
Active Military:	201,750	**Index of Martial Potency:**	4.35
Contiguous boundary:	Sudan 605 km	**Geopolitical Index:**	10

Eritrea is a small country which comprises what, before independence from Ethiopia in 1991, was the Ethiopian coastal strip on the Red Sea. In area, it is a small state but within the Rimland it is larger than Djibouti, Bulgaria and the three states of the Southern Caucasus. The population of almost five million is also small and exceeds only that of Djibouti, Armenia and Georgia within the Rimland. The GDP of Eritrea is approximately half that of Niger and Tajikistan and is under 20 per cent of that of Bahrain. Within the Rimland, only Djibouti has a lower GDP. The GDP per capita is equal to that of Djibouti, Ethiopia and Chad and, within the Rimland, is larger only than those of Niger and Somalia. However, Eritrea, as a result of the continuing boundary dispute with Ethiopia, maintains a military strength which is out of proportion to its other statistics. Furthermore, the military strength is a third larger again than that of Ethiopia. The only state in the Rimland with a higher figure for military strength is Pakistan. The Index of Martial Potency is rather lower than that for Ethiopia but among the higher figures for the Rimland. Eritrea has one long boundary in common with a Middle Eastern state, Sudan. Given the continuing problems of the boundary, the Geopolitical Index is 10.

Eritrea has very close links with the Middle East. The religion is Muslim and the language is Arabic. The trading pattern shows minor Middle Eastern links. Turkey (4.4 per cent) and Jordan (4.2 per cent) are import partners and Turkey supplies 4 per cent of exports. Like Pakistan, Eritrea has established a maritime boundary with a Middle Eastern state, Yemen.

The main international dispute concerns Ethiopia. In 2002 agreement was reached to accept the Ethiopia–Eritrea Boundary Commission's (EEBC) delimitation but neither party accepted the detailed revised line put forward in the EEBC demarcation statement (November 2006). The 25km-wide Temporary Security Zone in Eritrea has been monitored since 2000 by the UN Mission in Ethiopia and Eritrea (UNMEE). The other dispute is with Sudan over claims that Eritrea has supported Sudanese rebel groups.

No evidence has been put forward of drugs or people trafficking and there is not a significant number of refugees.

Ethiopia

Area:	1,127127 km²	**Comparison with Bahrain:**	1,694.9×
Population:	76,512 million	**Comparison with Bahrain:**	108×
GDP $:	74.88 billion	**Comparison with Bahrain:**	3.0×
GDP per capita $:	1,000		
Active Military:	152,500	**Index of Martial Potency:**	4.93
Contiguous boundary:	Sudan 1,606 km	**Geopolitical Index:**	7.5

Ethiopia is by far the largest of the Horn of Africa states, its size being exceeded in the Rimland only by Kazakhstan and the Sahelian states: Chad, Niger and Mali. Its population of 76.5 million is exceeded in the Rimland countries only by that of Pakistan. Its GDP is the fourth highest in the Rimland, exceeded only by those of Pakistan, Greece and Kazakhstan. However, in sharp contrast, its GDP per capita is one of the lowest in the Rimland, equal to those of Eritrea, Djibouti and Mali and exceeding only that of Niger and Somalia. The large size of its military puts it only behind Pakistan and Eritrea and its Index of Martial Potency puts it only behind Pakistan, Greece, Kazakhstan and Bulgaria. Ethiopia has a very long common boundary with Sudan which is relatively volatile, having a Geopolitical Index of 7.5.

Ethiopia has relatively strong Middle Eastern connections in that the Muslim population is put at 32.8 per cent. It also has trade relations with Saudi Arabia which provides 18.1 per cent of its imports and takes 5.8 per cent of its exports.

The major international issues in Ethiopia concern boundary disputes, refugees and drug trafficking. In 2002, Ethiopia and Eritrea decided to abide by the EEBC's delimitation decision but neither party has accepted the revised line produced in the EEBC statement of November 2006. The 25 km-wide Temporary Security Zone in Eritrea has been monitored since 2000 by UNMEE. There has been conflict with Somalia over rival claims within the Ogaden (Ethiopia) and the Oromo region (Somalia) and the Ethiopia forces invaded in January 2007. "Somaliland" secessionists provide port facilities in Berbera and trade ties to Ethiopia have helped overcome its landlocked location. The demarcation of the porous boundary to the west with Sudan continues to be hampered by civil unrest in eastern Sudan.

There are major groups of refugees in Ethiopia, primarily almost 74,000 from Sudan, 16,000 from Somalia and 10,700 from Eritrea. There are anything from 100,000 to 180,000 internally displaced persons as a result of the boundary war with Eritrea and ethnic clashes in Gambela. Ethiopia is a transit hub for heroin from south-west and south-east Asia destined for Europe together with cocaine trafficked to the markets of southern Africa. Qat is grown for local use and export principally to Djibouti and Somalia; money laundering is little developed.

Somalia

Area:	637,657 km²	**Comparison with Bahrain:**	958.9×
Population:	9.119 million	**Comparison with Bahrain:**	12.9×
GDP $:	5.259 billion	**Comparison with Bahrain:**	21.4%
GDP per capita $:	600		
Active Military:	None since 1999	**Index of Martial Potency:**	–
Contiguous boundaries:	–	**Geopolitical Index:**	–

In 1991 Somalia descended into turmoil, factional fighting and anarchy. In May of that year northern clans together declared an independent republic of Somaliland which has remained stable although unrecognised by any government. There is also a neighbouring self-declared autonomous state of Puntland, self-governing since 1998, but not aiming at independence. There remains a dispute over boundaries between Puntland and Somaliland. In the south, conditions have been particularly chaotic with famines and fighting. There was a two-year UN humanitarian effort but this was withdrawn in 1995. In October 2004 a Transitional Federal Government (TFG) was established but became deeply divided and by December 2006 controlled only the town of Baidoa. In June 2006 the Supreme Council of Islamic Courts (SCIC) defeated Mogadishu warlords to control the capital. They expanded south and threatened to overthrow the TFG. Consequently, Ethiopia invaded and with TFG forces drove the SCIC from power by the end of 2006, but fighting continues in south-west Somalia. As a result of all this, particularly the lack of a central representative government, Somalia has become the classic example of a failed state and is ranked third in the Failed State Index.

Of the states which are under a million square kilometres, Somalia is second only to Pakistan. It is a very large state with a population of just over nine million, a total which places it behind the very large states of the Sahel but ahead of three of the large Central Asian states. Of all the Rimland states, its GDP is higher only than those of Djibouti and Eritrea but its GDP per capita is the lowest at $600. There are no figures for military strength or Martial Potency as there has been no military since 1991.

Somalia is well integrated into the Middle East. The population includes 30,000 Arabs, the religion is totally Sunni Muslim and Arabic is one of the main languages spoken. The trading pattern of Somalia also shows strong links with the Middle East. The UAE (49.8 per cent), Yemen (21.5 per cent) and Oman (6 per cent) are all export partners while imports are provided by Oman (5.5 per cent), UAE (5.2 per cent) and Yemen (5 per cent).

Somalia is beset with disputes both internal and external. The governments of the "regional states", Somaliland and Puntland, seek internal support but have boundary claims on each other. There are also competing claims over Ethiopia's Ogaden and southern Somalia's Oromo region. Somaliland has gone the furthest towards secession by developing close relations with Ethiopia. It has provided port facilities at Berbera, vital for Ethiopia once it became landlocked. Somaliland has also established commercial ties with other regional states. There are said to be 400,000 internally displaced people.

SAHEL

The Sahel is the sub-Saharan region of Africa which stretches in a latitudinal belt across the continent and is best known for droughts and famine (Map 61). Of all the major areas in the world, the Sahel is arguably the one with the most extreme living conditions. It stretches from western Sudan through to the eastern part of Senegal and includes four countries which abut on to Middle Eastern states. Senegal, the state with only minor Sahelian influence on its eastern borderlands is a middle-range state in size but the other three are all extremely large and sparsely populated and are wholly Sahelian. The Sahel is a place with a range of extreme natural hazards, particularly recurring droughts, desertification, locust infestation and inadequate water. It is characterised by poverty, malnutrition and under-nutrition and nowhere is life more closely controlled by rainfall.

Major issues include the trafficking of people and drugs to the wealthier states of the Mediterranean coast and from there to Western Europe. Despite the obvious hazards, much of this trade is trans-Saharan. There is also economic migration on a large scale with the illegal movements of labour to the burgeoning markets of the Middle East and Europe.

Senegal

Area:	196,190 km²	**Comparison with Bahrain:**	295×
Population:	12.853 million	**Comparison with Bahrain:**	18.1×
GDP $:	20.6 billion	**Comparison with Bahrain:**	83.7%
GDP per capita $:	1,700		
Active Military:	13,620	**Index of Martial Potency:**	2.72
Contiguous boundary:	Mauritania 813 km	**Geopolitical Index:**	3.5

Senegal covers an area which is only 16 per cent of that of Niger, the smallest of its Sahelian neighbours. However, it is more extensive than the Rimland countries of south-east Europe, the southern Caucasus and Tajikistan. The population of 12.8 million is very much of the same order as that of the other Sahelian states, despite the fact that Senegal is so much smaller. Of the Rimland states only Pakistan, Ethiopia, Uzbekistan, Kazakhstan and, marginally, Niger have a higher population. Its GDP is the highest of the Sahel states and higher than the GDPs of Tajikistan and Kyrgyzstan. GDP per capita is low, but equal to that of Chad and far ahead of those of Mali and Niger. It exceeds the GDPs of all the states of the Horn of Africa. Senegal's military strength is low but in excess of the statistics for Mali and Niger while being only half the military strength of Chad. It nonetheless exceeds the military strength of Georgia, Tajikistan and Kyrgyzstan. The Index of Martial Potency is low but is very much in accordance with its military strength. Senegal has one common boundary with a Middle Eastern state, Mauritania and that is politically generally stable with a Geopolitical Index of 3.5.

The only obvious link between Senegal and the Middle East is the fact that 94 per cent of the population is Muslim. There are no Middle Eastern countries in its trading pattern.

The main issues concern boundary disputes, particularly separatist violence and transboundary raids together with arms smuggling from the Casamance region of Senegal to the Gambia and Guinea-Bissau. A resultant has been that some 16,000 people from Casamance have fled into the neighbouring countries while 2,500 from Guinea-Bissau have moved into Senegal to escape the conflict. There are nearly 20,000 refugees from Mauritania and there are said to be 22,400 internally displaced people. Senegal is a trans-shipment point for heroin from south-east and south-west Asia and cocaine from South America for Europe and North America.

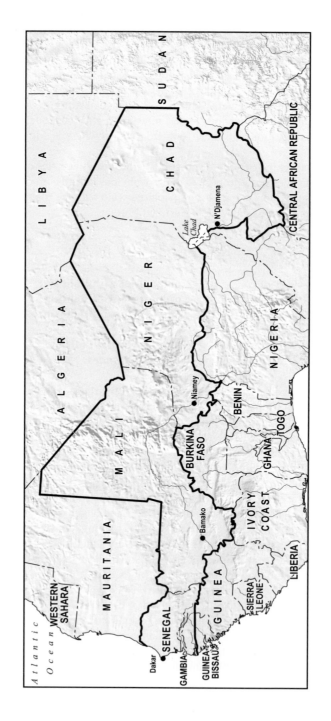

Map 61 Sahel

Mali

Area:	1.24 million km²	**Comparison with Bahrain:**	1,864.7×
Population:	12.324 million	**Comparison with Bahrain:**	17.4×
GDP $:	13.47 billion	**Comparison with Bahrain:**	54.7%
GDP per capita $:	1,000		
Active Military:	7,250	**Index of Martial Potency:**	2.81
Contiguous boundaries:	Algeria 1,376 km	**Geopolitical Index:**	0
	Mauritania 2,237 km		2.5

With an area of 1.24 million km², Mali is the smallest of the three main Sahel countries, all of which are among the largest states in the world. Its population is similar in size to other states of the Sahel and is marginally smaller than that of Senegal. The GDP is of a similar order to that of Niger but considerably smaller than the GDP of Senegal. At $31.5 billion, it exceeds, within the Rimland, the GDPs of Djibouti, Eritrea, Somalia, Tajikistan and Kyrgyzstan. However, the GDP per capita of $1,000 is equal to that of Djibouti which has just under 2 per cent of its area. The GDP per capita is larger only than those of Somalia and Niger. Mali's military strength is very low, lower than any state in the Rimland other than Somalia, that has no military, and Niger. The Index of Martial Potency is relatively high for the size of the military. Mali has two very long common boundaries with Middle Eastern states, Algeria and Mauritania, and both are largely inactive.

The clear connection with the Middle East is through religion since 90 per cent of the population is Muslim. Ethnically, 10 per cent of the population is listed as Tuareg and Moor, names which indicate a Middle Eastern affinity. The trading pattern shows no Middle Eastern partners.

Mali has no international disputes and is not involved in trafficking. The number of refugees is put at 6,300 from Mauritania.

Niger

Area:	1.267 million km²	**Comparison with Bahrain:**	1,905.3×
Population:	13.273 million	**Comparison with Bahrain:**	18.7×
GDP $:	8.902 billion	**Comparison with Bahrain:**	36.2%
GDP per capita $:	700		
Active Military:	5,300	**Index of Martial Potency:**	2.38
Contiguous boundaries:	Algeria 956 km	**Geopolitical Index:**	4
	Libya 354 km		0

Niger is a typically large Sahelian state, its area being between that of Mali and Chad. Apart from Chad, the only Rimland state that is larger is Kazakhstan. Its population of 13.2 million is the largest of the Sahelian Rimland states. It is exceeded in the Rimland only by the populations of Pakistan, Ethiopia, Uzbekistan and Kazakhstan. Its GDP of $8.9 billion is the lowest of the Sahelian states, only 43 per cent of that of Senegal, and within the Rimland, exceeds only those of Djibouti, Eritrea and Somalia. The GDP per capita is the lowest of every state in the Rimland except Somalia. The military strength of Niger is, apart from Somalia, the lowest of all the states of the Rimland and its Index of Martial Potency is also the lowest in the region other than that of Chad. Niger has two relatively long common boundaries with Middle Eastern countries, Algeria and Libya, and while there has been a little political activity along the former, the latter is quiescent. In common with the other states of the Sahel, the main link with the Middle East is the fact that a high proportion of the population, in this case 80 per cent, is Muslim. Ethnically, 9.3 per cent are listed as Tuareg.

The only international disputes concern boundaries and can be considered dormant or potential problems. Libya claims approximately 25,000 km² in the Tommo region. The boundary with Benin, including the tripoint with Nigeria remains undemarcated. The Lake Chad Commission's request to ratify the delimitation treaty has been addressed only by Nigeria and Cameroon so that the Niger–Chad and Niger–Nigeria boundaries are unratified.

There is no reported drug activities or people trafficking and the number of refugees is insignificant.

Chad

Area:	1.284 million km²	**Comparison with Bahrain:**	1,930.8×
Population:	10.111 million	**Comparison with Bahrain:**	14.3×
GDP $:	15.9 billion	**Comparison with Bahrain:**	64.6%
GDP per capita $:	1,700		
Active Military:	25,350	**Index of Martial Potency:**	2.16
Contiguous boundaries:	Libya 1,055 km	**Geopolitical Index:**	5
	Sudan 1,360 km		6

Chad is the largest of the four Sahel countries and its area is exceeded only by Kazakhstan within the Rimland. However, it has the lowest population of the Sahel countries, approximately three-quarters of that of the most populous, Niger. Its GDP considerably exceeds those of Niger and Mali and it has the highest GDP per capita, within the region, equal to that of Senegal.

Its military strength of just over 25,000 is low but is, by some distance, the highest in the Sahel. It is approximately twice that of Senegal, the next highest, and five times that of Niger, the lowest. In the remainder of the Rimland, the figure exceeds those for two Central Asian states and is virtually equal to that of Turkmenistan. Within the Southern Caucasus, it is more than twice the military strength of Georgia. However, the Index of Martial Potency is the lowest in the Rimland. Chad has two long boundaries in common with Middle Eastern states, Libya and Sudan, and both have been subject to some political turbulence being accorded Geopolitical Indices of 5 and 6 respectively.

Chad is by far the most closely attached to the Middle East of all the states of the Sahel. Of the population 12.3 per cent is ethnically Arab, 53.1 per cent is Muslim and, with French, Arab is an official language. Nonetheless, the trading pattern is little affected with no export partners and Saudi Arabia accounting for only 5 per cent of the imports.

The major cause of international disputes have been boundary disputes. Since 2003 the boundary with Sudan has been the scene of constant instability and hundreds of thousands of Darfur residents have been driven out of Sudan by the military and the Janjaweed armed militia into Chad. A relict of the Aozou Strip conflict is that Chadian Aozou rebels continue to reside in southern Libya. In common with Niger, Chad has not ratified the delimitation treaty, despite pressure from the Lake Chad Commission. In fact, only Nigeria and Cameroon have addressed the Commission's wishes. Therefore, the delimitation of the Chad–Niger boundary has yet to be ratified. Chad is ranked fifth on the Failed State Index.

No drug activity or people trafficking is reported. There are said to be 234,000 refugees from Sudan and 54,200 from the Central African Republic. Internally displaced persons number about 180,000.

Rimland seas

Water bodies are also of considerable significance in the Middle East. In many ways, the Gulf is perceived as the core of the Middle East. It is bounded by eight Middle Eastern littoral states. The Rimland is marked by five seas: the Mediterranean, the Black, the Caspian, the Arabian and the Red seas and characteristics of these influence the Middle East (Map 62).

The Mediterranean Sea

The Mediterranean is an extensive sea, marginal to the Atlantic Ocean and located at the centre of the World Island between Europe, Asia and Africa. The states of the southern littoral are all part of what is considered to be the Middle East. On the northern side of the Mediterranean, Turkey is the only Middle Eastern littoral state. The eastern side of the sea, the Levant coastline, comprises all Middle Eastern states. Of the two island states, Cyprus is also Middle Eastern.

The Mediterranean Sea has been crucial in the history of mankind from the earliest times to the present. It saw the growth of the great civilisations of Egypt, Greece and Rome and its importance has remained until today when it constitutes the southern flank of NATO and is patrolled by the US Mediterranean Fleet. Geopolitically, even following the opening of transatlantic trade, the Mediterranean has rarely been a backwater. For example, in 1565 Malta, in the centre of the Mediterranean, was the scene for the Great Siege by the might of Turkey. At stake was the Christian inheritance of Europe. In 1940–3 Malta was again under siege during what was the critical phase of the Second World War. Today, the east–west confrontation of the Cold War has been replaced by a less clearly defined north–south division of power. The north–south divide runs centrally down the Mediterranean. To the north is the European Union of more wealthy, developed countries. To the south is the developing world of Egypt, Libya and the *Maghrib*. At the eastern end, the long-running conflict between Israel and Palestine remains the dominant factor of Middle Eastern life.

The Mediterranean has eighteen littoral states together with the Gaza Strip. Of these states, Bosnia needs to be highlighted since it has a coastline of only 20 km on the Adriatic Sea. There are two island states: Malta and Cyprus. Of the eighteen littoral states, nine are Middle Eastern together with the Gaza Strip and Cyprus.

There are relatively few agreed maritime boundaries, the most complete delimitations being those of Italy and Tunisia (Map 62). The boundaries between Italy and the Balkan states (other than Albania), Greece, Tunisia, France (Corsica) and Spanish possessions have all been delimited. These boundaries are all between opposite states. The maritime boundary between two adjacent states, Tunisia and Libya, was produced as a result of an ICJ ruling as was the partial boundary between Libya and Malta. Elsewhere, the only boundaries are those between Cyprus and the sovereign base areas retained by the UK. The most intractable boundary, which has not even reached the stage of allocation, is that between Greece and Turkey in the Aegean Sea.

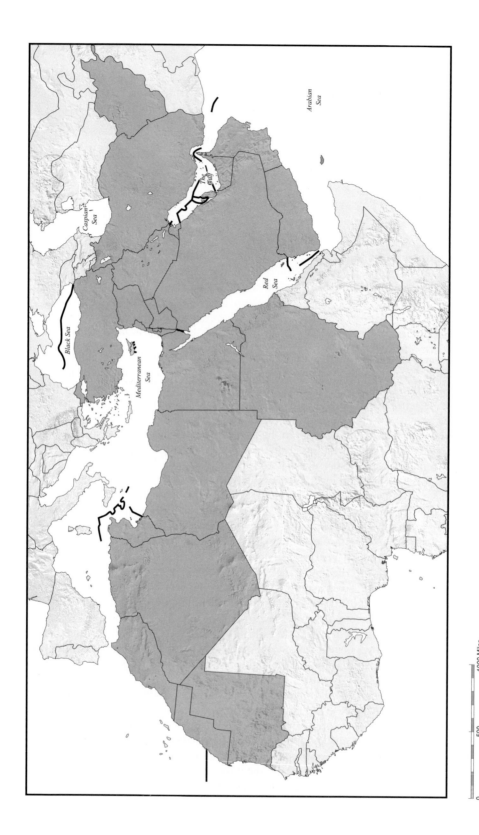

Map 62 Middle East: Rimland seas and maritime boundaries

Arabian
Sea

Caspian
Sea

The
Gulf

Red
Sea

Black Sea

Mediterranean
Sea

1000 Miles

500

1500 Kilometres

500   1000

0

0

Throughout history, the Mediterranean has hosted major maritime route ways but in 1869 with the opening of the Suez Canal the Mediterranean achieved a global importance for sea-lanes of communication which it has never lost. The sea-lane down the centre of the Mediterranean is particularly important as a tanker route from the Arabian Peninsula and, increasingly, the Black Sea to Western Europe and North America. The Mediterranean is also characterised by a number of choke points, narrows in which shipping is highly concentrated and therefore, in times of instability, open to attack. At the western end is the Strait of Gibraltar and at the eastern end the man-made choke point of the Suez Canal. In the central area, Malta lies in the narrowing of the route ways between Sicily and Tunisia. The most obvious marginal choke points are those that separate the Black Sea from the Mediterranean. The Turkish Straits, particularly the Bosporus, are the scene of major shipping concentrations, not only tankers transporting petroleum from the Caspian basin but also warships of the Russian Black Sea fleet.

The Black Sea

The Black Sea is a marginal sea of the Mediterranean. During the Cold War the littoral states, other than Turkey, were all members of the Warsaw Pact. This point emphasises the importance of Turkey to NATO during the whole of that period. Since the break up of the Soviet Union, Georgia and Ukraine emerged as separate states so that there are now six littoral states of which one, Turkey, is Middle Eastern.

There is one maritime boundary, agreed between Turkey and the Soviet Union (1973, 1978 and 1987). As a successor state, it is assumed that this boundary will constitute the delimitation between Georgia and Turkey but the position of Georgia on this issue is unknown. The boundary with Ukraine is partial. The Black Sea is a key route way as an outlet for southern Russia but more importantly the sea-lanes are crucial for the movement of petroleum from the Caspian Sea basin. However, much is already moved by pipeline and further additions to the network are planned. The Black Sea has become a vital transport node by both land and sea. The main limitation on sea-borne traffic is the capacity of the Turkish Straits, particularly the Bosporus. At its narrowest, the Bosporus channel is approximately 0.5 nml wide and has to cater for traffic flows into and out of the Black Sea but also for local traffic including ferries. It is the most congested choke point in the world, not in the number of ships per day, but in terms of shipping concentrations.

The Caspian Sea

The Caspian Sea is unconnected to the world ocean except through a canal system and this has given rise to detailed speculation, with legal implications, as to its status. There are five littoral states only one of which is Middle Eastern. If the Caspian is a sea, then the Law of Sea procedures should apply for boundary delimitation. If it is a lake, there are several possible approaches. There are over fifty major lakes on international boundaries from which a parallel might be chosen. Until there is agreed delimitation, exploitation of the seabed resources will be limited. As yet, there is no agreement on the water column and this could affect the fishing rights, particularly with regard to sturgeon. At present, equidistant seabed treaties have been ratified between Kazakhstan, Azerbaijan and Russia. The delimitation of a seabed boundary with Turkmenistan remains under discussion. Bearing in mind the size of its accepted area of seabed when the only negotiation required was with the Soviet Union, Iran has proposed a different procedure. Route ways in the Caspian are concerned primarily with petroleum.

The Arabian Sea

The Arabian Sea is a northern part of the Indian Ocean with two key marginal seas: the Gulf and the Red Sea. There are four littoral states of which two, Oman and Yemen, are Middle Eastern. The only agreed maritime boundary is a short length between Oman and Pakistan. The Arabian Sea is mainly significant for the petroleum and liquefied natural gas sea-lanes from the Gulf and the Red Sea. Trans-Arabian sea movement is concerned with migrants and economic refugees. Most legitimate labour movement is by air. Masirah Island (Oman) remains a US base but the former Soviet anchorage off Socotra Island (Yemen) is no longer in service.

The Gulf

The Gulf has no outlet at its northern end and is a vital sea-lane in the global petroleum network. Iran is the littoral state for effectively half the Gulf together with the Gulf of Oman. The remaining seven littoral states, also Middle Eastern, share the opposite coastline. Iraq has a coastline length of only 58 km. The other key point about the Gulf is the Strait of Hormuz which separates the Gulf from the Gulf of Oman. This is the best-known choke point in the world as a result of the passage of petroleum, and interdiction has been frequently threatened. Maritime delimitation is nearing completion in the Gulf. The maritime boundaries of Qatar and Bahrain were completed with the ICJ decision of 2001. Other maritime boundaries are discussed in Section C.

The Red Sea

Unlike the Gulf, the Red Sea has a northern outlet through the Suez Canal. It also has a significant marginal sea, the Gulf of Aqaba. The Red Sea together with the Gulf of Aqaba has eight littoral states, six of which are Middle Eastern. Djibouti is located at the entry point to the Red Sea and can be considered both a Red Sea and a Gulf of Aden state. The Gulf of Aden is the approach to the Red Sea which is located between Yemen and Somalia. The Gulf of Aqaba is important in that it allows Jordan to have a maritime outlet, albeit a coastal length of only 26 km. A maritime boundary was agreed between Jordan and Israel in 1996. In the Red Sea itself, the only maritime boundary between opposite states was agreed by Yemen and Eritrea. The only boundary between adjacent states is between Saudi Arabia and Yemen. There is a seabed resource agreement between Sudan and Saudi Arabia.

The key sea-lanes of the Red Sea are concerned with petroleum and their importance has been clearly illustrated when the Suez Canal has been closed and tankers have had to circumnavigate the Cape of Good Hope. The major choke point is Bab al Mandab at the southern end of the Red Sea but a second choke point, the Strait of Tiran, separates the Red Sea from the Gulf of Aqaba. Obviously, the Suez Canal constitutes a man-made choke point.

Section E
Key issues

This section provides overviews of the most significant geopolitical issues in the contemporary Middle East. The first part focuses on war and conflict in the region. The emphasis is on conflicts or issues that have contemporary relevance, including the 2006 war between Israel and Hezbollah because its reverberations are likely to be felt throughout the region for some time to come. Wars prior to the 1980s are not included (with the exception of Israeli–Arab wars). Also excluded are certain civil wars, such as the Lebanese civil war and the war between Hamas and Fatah. Though important, both were primarily urban wars that are neither easy, nor particularly useful to depict on maps. In addition to wars, the section also deals with unconventional weaponry (chemical, biological and nuclear) that either could be used in future Middle Eastern conflicts, or could be the source of conflict. For example, an Israeli/US strike on Iran's nuclear facilities, perhaps along the lines of Israel's bombing of Iraq's Osiraq nuclear reactor in 1981, can certainly not be ruled out in the short-term future.

The second part examines three political issues of particular importance in the contemporary Middle East. The success of the George W. Bush administration in achieving its stated goal of democratising the region can be judged against the evidence provided in the entry on democracy levels. The entries on the Kurds and the Shi'a reflect the elevated profile both communities have enjoyed since the ousting of Saddam Hussein. Though there are certainly other communities that could have been included for analysis, the Kurds and the Shi'a have the greatest potential to disturb the established order in the Middle East. The Shi'a revival, epitomised by the victory of Shi'a parties in the Iraqi elections, represents a serious challenge to established Sunni Arab powers in the Middle East. Many experts continue to downplay the significance of sectarian identity as a mobilising force, but it is difficult to dispute the fact that the Sunni/Shi'a divide is perhaps more relevant to Middle Eastern politics today than it has been for more than a thousand years. The Kurds are probably the only community with the capacity to threaten established boundaries in the region. The Kurds have traditionally failed to mobilise across boundaries and, as of 2009, the Kurds of Turkey, Iran and Syria, seem unlikely to challenge the integrity of their respective states' boundaries. The Kurds of Iraq are the exception. Though Iraqi Kurdish leaders have so far resisted popular pressure to push for independence, the ongoing struggle between Arabs and Kurds to define the nature of the new Iraqi state has great potential to precipitate the partition of Iraq. Certainly, if the Iraqi government is not prepared to compromise on core issues of concern to the Kurds, such as the status of Kirkuk, then Kurdish leaders may feel they have little to lose from secession.

The third part analyses some of the central issues at stake in the Israeli–Arab/Palestinian conflict. The two wars in which significant territories were acquired by Israel (1948 and 1967) are both included. This part also examines some of the most contentious issues associated with the Israeli–Palestinian conflict: settlements, checkpoints, the separation barrier and refugees; as well as providing an overview of the peace process at it unfolded during the 1990s.

The fourth and fifth parts deal with the two major ongoing conflicts in the region, Iraq and Afghanistan. The fourth part looks at three of the issues that have potential to create instability in the future in Iraq, while the fifth part offers a background perspective of how Afghanistan has reached its current state. It also deals with one of the most intractable of Afghanistan's (and the world's) current problems, namely the drug trade.

This section cannot attempt to provide an exhaustive account of all the issues in the contemporary Middle East. The focus is on those that have shaped, or will shape, the region's geopolitics over the next decades.

War and conflict

The Iran–Iraq War

The twentieth century's last great war of the masses pitted Iraq's oppressive, staunchly secular Sunni Arab-dominated regime against the fanatically religious Shi'a Persian regime of Iran. The culmination of centuries of Arab–Persian and Sunni–Shi'a mistrust and hostility, the war lasted for eight years and achieved no tangible gains for either side. It was an exercise in futility perpetuated by personal hostility between the two regime leaders, Saddam Hussein and Ayatollah Khomeini, that ended up costing hundreds of thousands of lives and tens of billions of dollars. The fallout from the war contributed directly to the subsequent Gulf War (Operation Desert Storm) of 1991.

The rise in tensions between Iran and Iraq during the 1970s played out against the backdrop of a Cold War Middle East in which the Shah's Iran played the role of reliable Gulf policeman for the USA, while Iraq moved closer toward the Soviet sphere of influence. Faced with a ruinous Kurdish rebellion in the north that was bankrolled by money and weapons from Iran, the USA, and Israel, in 1975 Saddam was forced to conclude the humiliating Algiers Agreement with the Shah that resolved the long-running Shatt al Arab boundary dispute in Iran's favour. However, the real trigger for war was the advent of the fundamentalist Shi'a regime in Iran following the 1979 revolution. The Iranian spiritual and political leader Ayatollah Khomeini poured fuel on the fire when he called upon Iraq's majority Shi'a population to rise up in rebellion against Iraq's secular, Sunni-dominated regime. To the Iraqi regime, which had struggled since the 1950s to suppress political Shi'ism as an evolving force, Khomeini's words constituted a grave threat to Iraq's political unity. After reciprocal shelling and a number of border skirmishes across the shared boundary throughout 1980, on 22 September, Iraq invaded Iran.

Saddam's motives for attacking are imperfectly understood. Most likely he planned for a lightening advance across the boundary and a swift land grab, including the ethnically Arab Iranian territory of Khuzistan. From there, it was possible that Iran's Arabs would join with their ethnic cohort and trigger a general uprising against the regime in Tehran. At a minimum, Iraq would be in a favourable position from which to reopen negotiations on the 1975 Agreement. This proved to be a serious miscalculation on the part of Saddam Hussein. Initially, Iraqi forces made swift advances into Iranian territory. By October Iraqi forces had captured Khorramshahr and were besieging Iran's second-largest city, Abadan. However, far from uniting with invading forces, Iranian Arabs rallied behind the regime in Tehran and offered stiff resistance. The extent of Saddam's miscalculation became apparent when the Iraqi advance stalled and efforts by Iraq to negotiate a diplomatic end to hostilities were rejected outright by Khomeini. Khomeini instead vowed to pursue the war until Saddam's regime had been destroyed.

The temporary stalemate lasted until mid-1982 when the Iranians launched their counterattack and recaptured Khorramshahr in a human wave assault that led to the capture of 15,000

Iraqis. To avoid a catastrophic defeat, Saddam sensibly withdrew his forces back across the boundary. In July 1982, the Iranian forces entered Iraqi territory for the first time and advanced to within ten miles of the southern port city of Basra. Efforts to capture Basra throughout 1983 were repelled by the Iraqis using chemical weapons, the first of many such uses during the eight-year conflict. Between 1984 and 1986, the only important territorial gain by either side was made by Iran when it captured the oil-rich Majnoon Islands. Otherwise, both sides concentrated on each other's oil-related infrastructure and, at times, key cities in an intensifying campaign of missile attacks.

A major offensive by Iranian forces in February 1986 resulted in the capture of large swathes of Iraqi territory on the Fao Peninsula and placed Basra directly in jeopardy. This development changed the complexion of the war entirely. Up to this point, the USA had supplied arms to both parties, but faced with the undesirable prospect of a complete Iranian victory, the USA began to intervene overtly on the Iraqi side. In response to a request from Kuwait, US naval vessels began to escort oil tankers to protect them against Iranian missile attack. With the assistance of chemical weapons and US satellite intelligence, Iraqi forces recaptured the Fao Peninsula in April 1988, and in May, retook the Majnoon Islands. After an escalating series of incidents between US naval forces and Iranian vessels, in July 1988, the USS *Vincennes* shot down an Iranian airliner killing all 290 on board. The USA claimed the incident was accidental (though it subsequently paid compensation to relatives of the survivors), but it was sufficient to convince the Iranians that the USA would never permit the total defeat of Iraq.

The eight-year Iran–Iraq War was a disaster for both sides. The Iraqis suffered somewhere in the region of 500,000 casualties and the Iranians up to one million. Added to its human losses, Iraq sustained damage to its infrastructure estimated at over $200 billion, and had been forced to borrow $100 billion to sustain its war effort, a crippling debt burden that had important repercussions after the war had finished.

Faced with economic disaster, Saddam Hussein pleaded with fellow Arab states, including Kuwait, to forgive Iraq's debts, and requested OPEC to reduce oil production to drive up the price and increase oil revenues flowing into Iraq's coffers. Kuwait refused to forgive the debt and persisted in exceeding its OPEC-approved quota. When Washington appeared to give the green light to an Iraqi attack on Kuwait, Iraq invaded in August 1990, precipitating the Gulf War of 1991.

The Iran–Iraq War also witnessed the most extensive use of chemical weapons on the battlefield since the First World War. Iraq used chemical weapons repeatedly against both Iranian forces and against Iraqi Kurds in the north, providing one of the key moral justifications for the US invasion of Iraq in 2003. The fact that Iraq had assembled its chemical and biological weapons arsenals from Western sources, including the UK, Germany and the USA, added irony to the justification. Indirectly, therefore, the Iran–Iraq War and its aftermath led to both subsequent US attacks on Iraq.

In the end, the war was a pointless exercise. Though Iraq claimed victory, Map 63 illustrates the fact that territorially the war changed nothing. The boundary between the two countries at war's end was almost exactly what it had been before the war started.

The 1991 Gulf War and aftermath

The roots of the 1991 Gulf War can be traced directly to the aftermath of the Iran–Iraq War. Though Iraq has a long-standing territorial claim on Kuwait, the precipitate causes of the conflict stemmed from the dire economic condition in which Iraq emerged from that war. Iraq had been forced to borrow heavily to maintain its war-fighting capacity while simultaneously preserving

Map 63 Iran–Iraq War

the standard of living of Iraqi civilians. In total, Iraq borrowed up to $100 billion, with the majority coming from Saudi Arabia and Kuwait. At the end of the war, with Iraq's economic infrastructure in ruins, and with a million-man army to fund, Saddam Hussein tried unsuccessfully to convince Kuwait to write off its debt.

In 1990, Iraq's oil revenues amounted to only $13 billion, barely sufficient to pay the army, let alone make the $200 billion investment to rebuild the country's shattered infrastructure. In a desperate effort to increase oil revenues, Iraq attempted to convince OPEC to reduce oil production. Kuwait continued to produce at above quota levels, and worse, reportedly began to pump oil out of the Rumaila oilfield that straddles the Iraq–Kuwait boundary. During the summer of 1990, as Iraqi troops began to mass along the Kuwaiti boundary, US officials made a series of statements that strongly implied that the USA would take no action in the event of an Iraqi assault on Kuwait. After final mediation efforts failed, Iraq invaded Kuwait on 2 August 1990.

A shocked Saudi Arabia immediately requested military assistance from the USA, and President George Bush duly obliged, dispatching 40,000 troops in Operation Desert Shield. Saddam Hussein responded by formally declaring Kuwait to be the "nineteenth province of Iraq" and by mid-August 1990, it became clear that confrontation was likely.

At the UN, the USA used its diplomatic muscle to secure passage of Security Council Resolution 678 which authorised member states to use "all necessary means" to drive Iraqi forces out of Kuwait unless Iraq withdrew by 15 January 1991. In the meantime, the USA mobilised a coalition force comprising some 600,000 troops from over forty countries. When the deadline expired without an Iraqi withdrawal, Operation Desert Storm was launched on 17 January 2003. The first stage of the assault was a massive aerial bombardment that lasted for six weeks and destroyed much of Iraq's military and civilian infrastructure. Subsequently, the brief land war was a rout of embarrassing proportions for Iraqi forces.

Though Iraq had one of the larger armies in world the morale of troops was exceptionally low and most surrendered at the first sight of coalition forces. Those that opted to fight were decimated by the coalition's technologically vastly superior weaponry. Within forty-eight hours of the commencement of the ground invasion, all Iraqi forces had been driven out of Kuwait, and subsequently, the only decision that remained for US generals was how great a defeat to inflict upon Iraq's by now defenceless army. Ultimately, as the war had degenerated into the senseless slaughter of unprotected Shi'a and Kurdish conscript troops, the USA announced a ceasefire and offered surrender terms to Iraq.

On the one hand, the war was a comprehensive defeat for Iraq; on the other, it was an incomplete victory for the USA and its coalition allies. Thirteen days before the ceasefire was announced, President Bush called publicly for the people of Iraq to "take matters into their own hands" by rising up against the regime of Saddam Hussein. This was interpreted by Iraqis as an offer of military support for any such undertaking. On 28 February 1991, a large-scale uprising against government forces broke out among the Shi'a population in the south, to be joined a few days later by a better organised Kurdish uprising in the north. By 19 March, sixteen of Iraq's eighteen provinces were in open revolt. Unfortunately for the rebels, the ceasefire had been called before Saddam Hussein's most loyal and effective military units, the Republican Guard, had been destroyed. Also, as part of the surrender negotiations, Iraq had requested, and obtained, the right to continue to use helicopters over Iraqi airspace. When President Bush's inaction made clear that his support for the rebels was entirely rhetorical, Saddam Hussein used a combination of Republican Guard units in the south and helicopter gunships in the north to suppress both rebellions in short order. By the end of March, the southern uprising had been quashed, while in the north the savagery of the attacks against the Kurds precipitated a mass exodus of Kurdish civilians from urban areas. By the end of April, approximately one million Kurds were in danger of either freezing or starving to death on the Turkish border.

In response, the UN issued a resolution calling for an end to the repression of civilians and the USA, UK and France announced the establishment of a no-fly zone for Iraqi aircraft north of the 36th parallel (Map 64). Under cover of the northern no-fly zone, and the implicit protection of coalition air power, the Kurds were able to return to their homes. Subsequently, they took advantage of their virtual independence from Iraq to establish the institutions of government and governed themselves with some success during the 1990s. The southern no-fly zone was established in 1992, nominally to protect the population of the south from further repression. Neither no-fly zone was ever approved by the UN, and so both were technically in violation of international law. As the 1990s progressed, the two no-fly zones increasingly lost any protective function and became, instead, a cover for the aircraft of the USA, UK and France (until 1998) to conduct offensive operations against Iraqi military facilities. Any pretence that the function of the no-fly zones was to protect the population was effectively shed in the run-up to the 2003 war

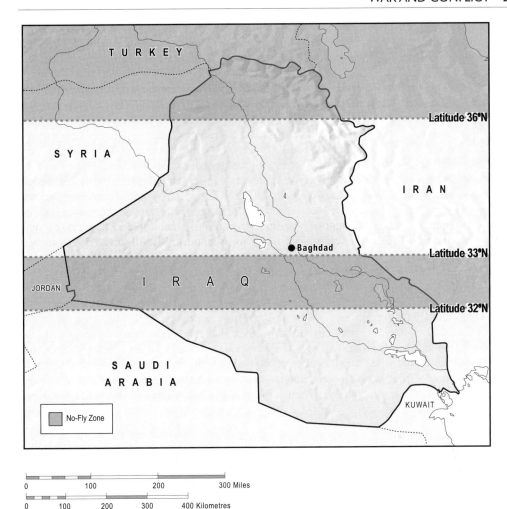

Map 64 Post-Gulf War (1991) no-fly zones

when the USA and UK launched a series of major air attacks on Iraqi air defence systems in parts of the south devoid of population

Operation Iraqi Freedom

In most respects the US- and UK-led war against Iraq in 2003 was a continuation of the 1991 Gulf War. Throughout the 1990s, the USA and UK launched daily air attacks on strategic targets in Iraq under cover of the two no-fly zones, and in this sense the war never really ended. The core of the conflict revolved around the status of Iraq's WMD programmes. A series of UN resolutions passed in the early 1990s had imposed comprehensive economic sanctions on Iraq and required Iraq to accept the destruction of all elements of its chemical, biological and nuclear weapons programmes. To supervise the destruction of these and to verify compliance, the UN established a UN Special Commission (UNSCOM), which began inspections of suspected Iraqi facilities shortly after the end of the 1991 war. A game of cat and mouse ensued whereby Iraqi officials would be instructed to reveal certain aspects of the country's WMD programmes, while systematically concealing others. The game was up when Saddam Hussein's son-in-law Hussein Kamel defected to Jordan and revealed the extent of Iraq's biological weapons programme and the elaborate plans in place to deceive UNSCOM inspectors.

Meanwhile, the most comprehensive sanctions programme ever inflicted on a country by the international community took a devastating toll on the Iraqi people, destroyed Iraq's once proud middle class and, by some estimates, led directly to the deaths of over 500,000 Iraqi children. The fundamental flaw with the sanctions policy was that both the Clinton administration and the Bush administration that preceded it had made clear that even the complete, verified disarmament of Iraq would not lead to a lifting of the sanctions. The sanctions would stay until Saddam Hussein was removed from power. Naturally, this did not furnish Saddam with much of an incentive to comply with inspections. Finally, in 1998 Iraq refused any further cooperation with inspections, UNSCOM inspectors were withdrawn from Iraq, and the USA and UK launched Operation Desert Fox, a 70-hour bombing campaign that achieved nothing of note. When UNSCOM departed Iraq, inspectors estimated that they had successfully destroyed 95 per cent of Iraq's WMDs.

WMDs subsequently became the stated rationale for the Bush administration's decision to launch Operation Iraqi Freedom to remove Saddam from power in 2003. Following the Twin Towers (World Trade Center) bombing of September 2001, the administration identified Iraq (along with Iran and North Korea) as part of an "axis of evil" and the race to war had begun. Citing "conclusive" intelligence evidence that Iraq maintained functioning WMD programmes, President George W. Bush and UK Prime Minister Tony Blair pushed the issue of Iraq's disarmament to the UN Security Council. When it became clear that France, and perhaps Russia, would veto any resolution authorising the use of force in the event of Iraq's verified disarmament, the USA and UK decided to act without UN approval. By any objective assessment, therefore, the ensuing war was a violation of international law.

When, unsurprisingly, Saddam Hussein refused to comply with President Bush's 48-hour deadline to leave Iraq, Operation Iraqi Freedom was launched on 20 March 2003. Unlike the 1991 Gulf War, the USA opted to invade with a much smaller, more mobile force and conducted only a very limited air war in preparation for the ground war. The battle plan (Map 65) involved US infantry and marines in a two-pronged, lightening dash for Baghdad, while British forces headed for Basra. A simultaneous attack from the north launched from Turkey was planned, but had to be abandoned shortly before the war after the use of Turkish bases was vetoed by Turkey's parliament. Meeting minimal resistance, US forces reached Baghdad in early April to find that Iraqi government officials had disappeared along with Iraq's armed forces, including the vaunted Republican Guard. On 9 April 2003, twenty-one days after the start of hostilities, Operation Iraqi Freedom was over.

Unfortunately, the legacy of the battle plan lives on to haunt Iraq. The preference of Defense Secretary Donald Rumsfeld for a small, highly mobile force was tactically astute, but strategically disastrous. A force of this size could not hope to occupy and control cities during the drive to Baghdad, so the decision was made to skirt most major southern cities. This left a security vacuum to be filled by religious Shi'a forces backed by their respective militias and in the process allowed these groups to become deeply rooted in the south. In the north, the absence of a second front meant the USA had to rely on Kurdish forces to take the major northern cities of Mosul and Kirkuk. Subsequently, the Kurds have retained a military presence in both cities which has created serious tensions in both with Arab and Turkoman inhabitants. The absence of sufficient US military force to provide security in Baghdad was, arguably, an even bigger

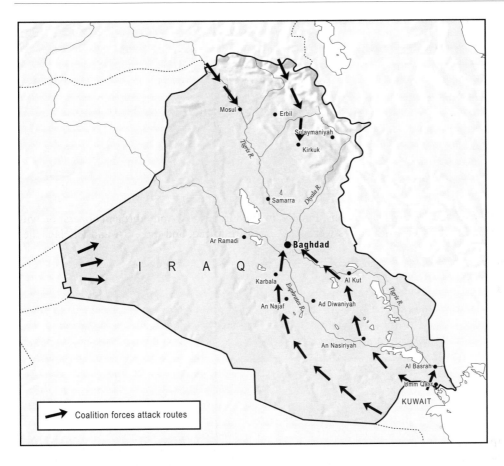

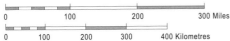

Map 65 Operation Iraqi Freedom

problem. The fall of Saddam's regime prompted a two-week orgy of looting that stripped the city's infrastructure bare and greatly hampered prospects for speedy reconstruction. Fundamentally, and against the advice of several senior military officers, Rumsfeld chose the force level adequate for the battle plan, but totally inadequate for a long-term occupation. That there are still 130,000 US troops in Iraq is ample testimony to the failure of the vision underlying Operation Iraqi Freedom.

The 2006 Hezbollah–Israeli war

The 33-day war between the IDF and the military wing of Hezbollah that took place in July–August 2006 was the latest phase in a conflict that dates back to the early 1980s (Map 66). As a distinct entity, Hezbollah was officially formed somewhere around 1982, and gained prominence

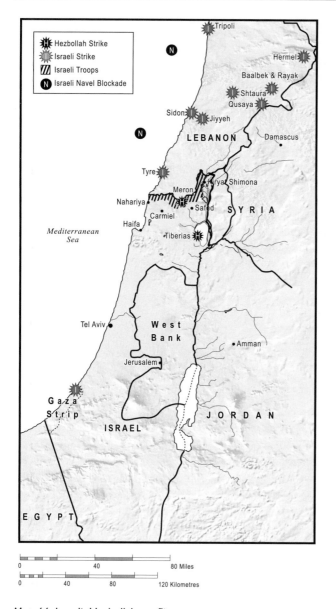

Map 66 Israeli–Hezbollah conflict

largely as a result of its resistance to Israel's 1982 invasion of Lebanon. The invasion came in the midst of the Lebanese civil war and was ostensibly designed to provide support for various Israeli-backed Lebanese militias to root out the PLO from its base in Beirut. The IDF retained a military presence in southern Lebanon until their final withdrawal in May 2000, but mainly relied on a proxy force, the South Lebanon Army (SLA), to confront Hezbollah and provide security around the Israeli–Lebanon boundary. Hezbollah, meanwhile, evolved into a multifaceted organisation that began participating in elections for the Lebanese parliament in the early 1990s while enhancing its military capabilities in the struggle against occupying Israeli forces.

During the 1980s, Hezbollah waged a low-level, but consistent and effective guerilla war against the Israelis and their proxies in southern Lebanon. During this period, Hezbollah honed its tactics of asymmetrical warfare, including assassinations and suicide bombings, but large-scale confrontations were infrequent. Periodically, the Israelis conducted more intense military operations in response to Hezbollah attacks. Hence, in 1993, the IDF launched Operation Accountability in response to a Hezbollah attack that killed five Israeli troops. The operation involved a sustained bombing of buildings and infrastructure in southern Lebanon that killed over one hundred Lebanese.

Then in April 1996, in response to Hezbollah rocket fire into Israel, the IDF mounted a large-scale bombing campaign (Operation Grapes of Wrath) that culminated in the deaths of over one hundred Lebanese civilians in the UN compound at Qana. The resulting international outcry over the incident (which Israel claimed was accidental), the virtual elimination of the SLA by Hezbollah, and the constant drain on resources required to sustain operations in Lebanon led the Israeli government to announce its withdrawal from Lebanon in 1998. The full withdrawal was completed in May 2000, thus bringing Israel into compliance with UN Security Council Resolution 425 of 1978. The Israeli withdrawal greatly enhanced the prestige of Hezbollah both within Lebanon and throughout the Middle East. Claiming credit for driving out the Israelis, Hezbollah emerged from the conflict as the most powerful political and military force in Lebanon.

Despite sporadic, low-level confrontations between Hezbollah and Israeli forces across the shared boundary, the war that erupted in July 2006 was unexpected. The immediate trigger was a cross-border raid, accompanied by rocket fire, by Hezbollah that resulted in the deaths of three Israeli troops and the kidnapping of another two. The intent was clearly for Hezbollah to capture IDF troops in order to conduct a prisoner swap, a fairly common practice between the two sides, but the audacity of the attack combined with the rocket fire provoked a stiff response from the Israeli government. Israeli Prime Minister Ehud Olmert described the attack as an act of war for which the Lebanese government must be held accountable because the government had failed to disarm Hezbollah (as required by UN Security Council resolutions) and because the attack had been launched from Lebanese soil.

Though subsequently claiming that the war was not against the Lebanese government or the people of Lebanon, Israeli forces began their assault with an air campaign that targeted civilian infrastructure deep inside Lebanon, such as the international airport in Beirut and the road linking Beirut to Damascus. Simultaneously, the Israelis imposed an air and naval blockade on Lebanon in an effort to prevent the resupply of Hezbollah forces. As Israeli air strikes continued to pound Hezbollah positions in the south and infrastructure targets in the north, Israeli ground troops entered Lebanon on 23 July 2006 and engaged Hezbollah fighters in a series of small-scale battles during late July and into early August. The stubborn resistance of Hezbollah's forces, who proved to be well trained and highly disciplined, meant that the outcome of these skirmishes, though inconclusive, were perceived by many in the region as a *de facto* victory for Hezbollah.

In all, approximately 120 IDF troops were killed during these engagements, while the number of Hezbollah dead was considerably higher. Estimates of Hezbollah dead ranged from 250 (Hezbollah's own claim) to 600 (Israel's claim). With much of Lebanon's infrastructure in ruins, and with a civilian death toll of over 1,000 on the Lebanese side and forty-three on the Israeli side, it was not clear that the war constituted a clear military victory for either side. One of the main aims of the Israeli action had been to eliminate Hezbollah's capacity to fire rockets at Israeli population centres, but throughout the duration of the conflict, Hezbollah continued to rain down rockets on targets in Israel. Moreover, the fact that Hezbollah had successfully resisted the might of the IDF, the first Arab army to do this since the creation of the state of Israel, meant that the war gave Hezbollah and its charismatic leader Hassan Nasrallah, a major propaganda victory. Subsequently, on the basis of its strengthened political position, Hezbollah was able to obtain *de facto* veto power in Lebanon's parliament and more than one-third of the cabinet positions in Lebanon's government of national unity announced in July 2008. In this sense the war was a significant political victory for Hezbollah, and a major political setback for Israel in that it boosted support for one of its most implacable and dangerous enemies in the region.

America's involvement in the Middle East

America's involvement in the Middle East has traditionally been driven by the perceived need to control access to the world's most extensive reserves of fossil fuels. US oil companies have been involved in the discovery and exploitation of the region's oil since the 1920s. Towards the end of the 1930s, US oil companies discovered oil in Saudi Arabia, Kuwait and Bahrain, and by the end of the Second World War, it had become apparent that the region possessed what one American official described as "one of the greatest material prizes in world history". During the early stages of the Cold War, US policy was geared towards preventing the Soviet Union from expanding its influence in the region, and displacing European powers (principally France and Britain) as the region's key power broker.

A seminal event occurred in 1953 when the USA and Britain responded to Iranian Prime Minister Mohammed Mossadegh's nationalisation of Iranian oil industry by orchestrating a coup (Operation Ajax) to remove him from power. Subsequently under the Shah, Iran was to remain one of America's most reliable allies in the region until the Islamic Revolution of 1979. In 1956, after Egypt's Arab nationalist leader Gamal Nasser nationalised the Suez Canal, the USA opposed the joint French–British–Israeli military campaign to retake control of the canal. The failure of the campaign to achieve its goals spelt the end of significant European involvement in the Middle East and paved the way for US dominance over the region. The US position during the Suez Crisis was anomalous in the context of its broader relationship with Israel. A staunch advocate of the creation of the Israeli state during 1947–8, the USA began to expand its strategic relationship with Israel during the Kennedy administration as a reliable anti-Soviet, and anti-Arab, nationalist force in the region.

Then, during the 1967 Arab–Israeli war, the USA provided diplomatic and material support to Israel, and was even accused by Arab states of providing tactical air support for Israeli forces (though this was never proved). During the 1973 Yom Kippur war, US assistance to Israel was more substantial and included airlifts of military equipment. It was also during the 1960s and 1970s that the USA started to provide loans and grants to Israel on a serious scale. Military grants to Israel began in 1974 and had, by 1979, reached $2–3 billion. Part of the reason for this huge expansion of aid from the USA to Israel was the signing of the Camp David peace agreement between Israel and Egypt. Camp David required Egypt to recognise the state of Israel officially, in return for which Israel withdrew from Egyptian territory occupied during the war. To smooth the deal, the USA agreed to provide large quantities of aid annually to both sides. As of 2009, Egypt and Israel remain the top two recipients of US foreign aid.

In 1979, the Islamic Revolution in Iran took US officials by surprise. At a stroke, the bedrock of the US alliance structure in the Gulf was shattered. During the Iran–Iraq War, the US strategy was neatly encapsulated in Henry Kissinger's immortal words, "too bad both sides can't lose". During the conflict, the USA provided funding and intelligence information to the Iraqis while covertly dealing arms to Iran. As a direct consequence of the debts incurred during the Iran–Iraq War, Iraq invaded Kuwait in August 1990. In response, the USA deployed troops to Saudi Arabia (Operation Desert Shield), to protect against a hypothesised invasion by Iraqi forces. The presence of US troops in the country charged with guardianship over Islam's holiest cities (Medina and Mecca) was perceived as an intolerable insult by many in the Muslim world, including Osama bin Laden.

Throughout the 1990s, the presence of US troops on holy ground was the major rallying cry of Islamists throughout the Middle East. Following the defeat of Iraq, and the imposition of the most stringent sanctions regime in modern history on its population, anti-American sentiment in the region escalated. The failure of the US-backed Israeli–Palestinian peace process during the 1990s, attributed by most in the Middle East to US bias in favour of Israel, further diminished the reputation of the USA. These three issues, the US military presence in Saudi Arabia, the US pro-Israeli position and the sanctions on Iraq, were cited by Osama bin Laden in his "Declaration of Jihad" issued in 1998. The end result of these grievances was the strikes of 11 September 2001.

The response of the George W. Bush administration to the events of 11 September was to target the Taliban regime in Afghanistan (as the protector of Al Qaeda), and then to move against the Ba'athist regime in Baghdad. The stated purpose of Operation Iraqi Freedom was to disarm Iraq of WMD, and the failure to find evidence of WMD led the USA to re-evaluate its goals. President Bush outlined the contours of a much more ambitious US policy for remaking the region along liberal democratic, free market lines, with Iraq's nascent democracy providing the shining beacon for other oppressed peoples in the region to follow. The deeper purpose was to address the root causes of terrorism in the region by eliminating the conditions (poverty and political oppression) that had generated such intense hostility towards the USA and its allies in the Middle East. Iraq's descent into an anarchic war of all against all has starkly illustrated the limitations of the new Bush doctrine. As of 2009, Iraq has a corrupt and deeply flawed democratic structure that sits atop a failed state, scarcely a compelling advertisement for liberal democracy.

Afghanistan is arguably in a worse condition. The Afghan government, though democratically elected, is deeply corrupt and heavily implicated in the drug trade. Its authority barely extends beyond the boundaries of Kabul, and the Taliban is now a major presence in over 50 per cent of Afghanistan. Elsewhere, the Bush administration's campaign to democratise the region has run up against an obvious, but potentially fatal problem. Aside from Israel and Turkey, all the region's leaders on whom the USA has built its current alliance structure are authoritarian and, with the possible exception of King Abdullah of Jordan, unlikely to prevail in free and fair elections. If the populations of the region are given the power to elect their own governments, the likely outcome is governments throughout the region that are significantly more anti-US and anti-Israeli than they are at present. The election of Hamas in the Palestinian parliamentary election of 2006 illustrates this point well.

More broadly, there is an evident, perhaps irreconcilable, tension between the US goal of securing the flow of oil from the Middle East and US staunch support for Israel. The USA is still pivotal to a resolution of the Israeli–Palestinian conflict, not because it is perceived as an "honest broker" but because only the USA has the leverage over Israel to extract the painful concessions that will be necessary on the Israeli side. Until the point comes that the USA is prepared to exercise its influence over Israel, any effort by the USA to promote democracy in the region is doomed to failure.

As of 2009, the USA finds itself more directly involved in the Middle East than at any time in its history. The US military bases in Saudi Arabia that alienated many in the region have been replaced by an expanding network of bases spread across the Middle East. The USA now has a military presence in virtually every Middle Eastern state, with the obvious exception of Iran (Map 67). When combined with the major US deployments in Iraq and Afghanistan, Iran is now effectively surrounded by US military power. At the same time, it is not clear that the enhanced military profile of the USA in the region has strengthened its strategic position. The USA finds

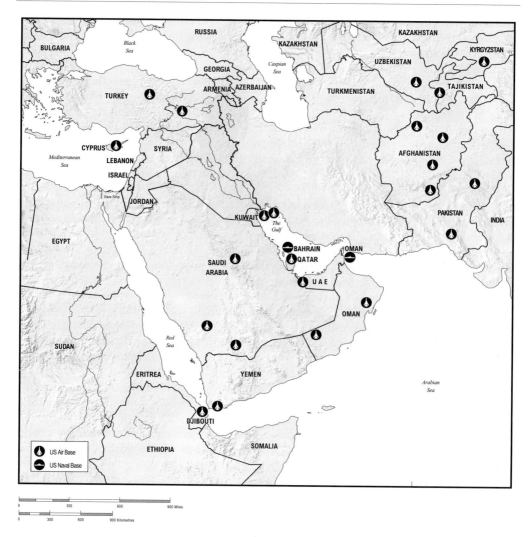

Map 67 US military bases

itself trying to isolate Iran and decrease its influence in Iraq, while simultaneously backing a government in Iraq that is dominated by Shi'a religious parties with strong ties to Iran. This symbolises the complexity of the task facing the USA as it struggles to deal with a region that is among the most geopolitically complex on Earth.

Weapons of mass destruction in the Middle East

The two most controversial WMD-related issues in the Middle East, Israel's WMD programmes and Iran's nuclear programme, are dealt with later in this section. There are, however, a number of states in the region that are suspected of possessing chemical and biological weapons programmes, and a larger number that possess ballistic missile capabilities (Map 68). The best

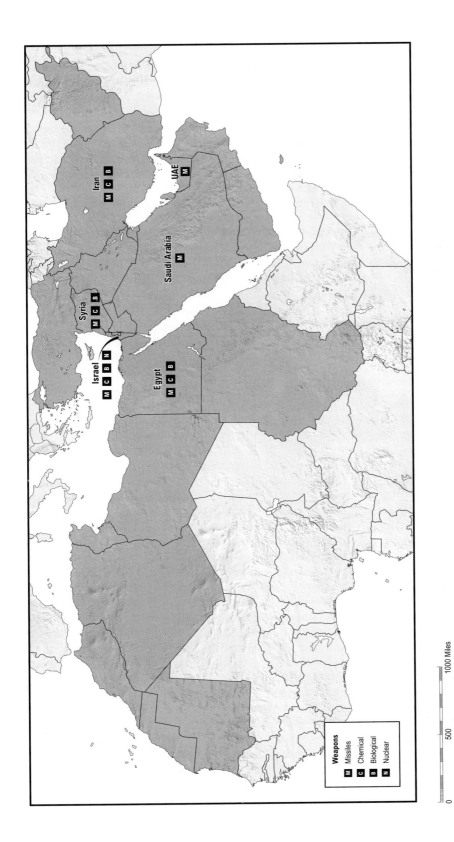

Map 68 Middle East: weapons of mass destruction

Weapons
M Missiles
C Chemical
B Biological
N Nuclear

Iran M C B
UAE M
Saudi Arabia M
Syria M C B
Israel M C B N
Egypt M C B

1000 Miles
500
1500 Kilometres
500 1000
0
0

documented chem/bio programmes in the region belonged to Iraq and Libya, states which have since either voluntarily renounced WMDs (Libya) or had their programmes forcibly dismantled by the UN (Iraq).

Until its catastrophic defeat in the 1991 war, Iraq had the most active and extensive WMD programmes in the region. Begun during the Iran–Iraq War, Iraq's chemical weapons programme produced vast quantities of mustard gas and the sophisticated nerve agents: VX, sarin and tabun. Chemical munitions were used by the Iraqis on multiple occasions during the war and proved highly effective against Iranian human wave attacks. They were then used in large quantities against Iraqi Kurds during the infamous Anfal campaign in 1988. Iraq also developed an advanced biological programme that included the weaponisation of anthrax and various toxins, such as botulinum.

Iraq's nuclear programme began in the early 1980s with the construction of a nuclear reactor at Osiraq, but received a serious setback when the reactor was bombed by the Israelis in July 1981. Subsequently, the Iraqi programme retreated underground and a covert uranium-enrichment programme was believed to be within months of producing the materials for a nuclear weapon when Iraq invaded Kuwait. After Iraq's defeat in 1991, however, Iraq's weapons and its WMD-related infrastructure were comprehensively dismantled by UN weapons inspectors, though not, apparently, to the satisfaction of the Central Intelligence Agency (CIA).

Libya's WMD programmes were far less extensive. The best developed was Libya's chemical weapons infrastructure. Prior to 1990, Libya is now known to have produced large quantities of mustard gas and sarin. In 2003, Libya renounced all of its WMD programmes and agreed to destroy all past stocks of chemical agents and convert its main chemical weapons facility to the production of pharmaceuticals. When Libya allowed US and British inspectors into Libya to examine its WMD programmes, inspectors found no evidence of a biological weapons capability, while Libya's nuclear weapons programme was found to be embryonic at best.

With both Iraq and Libya disarmed of WMD, what remain in the Middle East are those states suspected of possessing chemical or biological weapons. Iran certainly has produced a number of chemical agents in the past, including mustard gas and, perhaps, nerve agents. According to some experts, Iran actually used chemical weapons during the Iran–Iraq War. When Iran acceded to the Chemical Weapons Convention in 1998, it denied an ongoing programme, claiming it had dismantled its chemical infrastructure at the end of the war with Iraq. Many experts (including the CIA) doubt these claims, however, and believe that Iran has continued its programme with help from China and Russia. Iran is also suspected of pursuing a biological weapons programme, though little evidence has emerged of its extent or even existence.

Egypt has long been suspected of having an extensive and active chemical weapons programme. There is considerable evidence that, during the 1960s, Egypt actually used several types of chemical agents against Yemeni Royalist forces in support of South Yemen during the Yemeni civil war (1963–7). Subsequently, experts have described it as "almost certain" that Egypt continues to possess large stockpiles of chemical agents, including mustard gas, phosgene and various nerve agents. Egypt's biological weapons programme, if it exists, is shrouded in secrecy. US intelligence experts claim that Egypt had developed biological agents as early as 1972. Reports suggest that Egypt has worked on plague, botulism toxin and the encephalitis virus as bioweapons, but once again, the information is largely anecdotal. Though a signatory to the Biological Weapons Convention, Egypt has yet to ratify the treaty and refuses to do so until Israel gives up its WMD programmes. Egypt is not yet a signatory to the Chemical Weapons Convention. Syria is also believed, with some degree of certainty, to possess various chemical agents, including sarin, and is reportedly working to develop VX nerve agent. Syria is also accused of actively seeking a biological weapons capability, though its efforts are believed to be at the research and development (R&D), rather than weaponisation stage.

Among those states in the region that are known to possess a ballistic missile capability, Israel has by far the most powerful and capable missile force. Israel's operational Jericho-2 missile has a range of 1,500 km and is capable of carrying a nuclear warhead. When the Jericho-3 missile becomes operational, it will have a range of between 3,000 and 6,000 km, giving Israel a missile capability that exceeds those of all other states bar the acknowledged nuclear five. Saudi Arabia's Dong-Feng-3 missile, obtained from China, is the region's second-longest range missile, with a range of 2,600 km. Iran possesses various short range (<1,000 km) ballistic missiles including Scud-Cs and Scud-Ds and is also in the process of developing its Shahab-4 missile. Expected to have a range of over 2,000 km, the Shahab-4 will become the third-longest range missile in the Middle East when it becomes operational. Other Middle Eastern states with ballistic missile capability include Egypt (Scud-B and Scud-C), Libya (Scud-B), Syria (Scuds B, C and D), the UAE (Scud-B) and Yemen (Scud variants). Turkey also possesses various missiles of limited range that would qualify as ballistic missiles.

The two major problems the Middle East faces with respect to WMD are, first, Israel's unacknowledged possession of nuclear weapons, which provides a powerful incentive for all states in the region to arm themselves with unconventional weapons for deterrence purposes; and second, the evolving Iranian nuclear programme. If Iran's programme turns out to be geared towards acquiring a nuclear weapons capability, it runs the risk of triggering a proliferation spiral in the region that could result in multiple, nuclear-armed states facing each other in the Middle East.

Israel's weapons of mass destruction

Israel is the only state in the Middle East known to possess nuclear weapons. Although Israel has never definitively tested a nuclear device, and maintains an official position of nuclear ambiguity, there is a high degree of consensus among experts that Israel possesses somewhere between 100 and 300 usable nuclear devices. As one expert describes it, the existence of the Israeli bomb is "the world's worst kept secret". The Israeli nuclear programme started in the mid-1950s and emerged as a consequence of close cooperation with the French government. The French, indeed, supplied the Israelis with the plutonium-producing nuclear reactor at Dimona that remains at the heart of their nuclear weapons programme. A French company also constructed several key facilities at Dimona, including an installation for extracting usable plutonium from the reactor's spent fuel. Data from nuclear tests are generally believed to have been supplied by both France and the USA. The Dimona reactor became operational in 1964, and by 1966, Israel, again with French assistance, had completed the R&D stage of its programme and possessed sufficient plutonium to fashion a small number of rudimentary nuclear weapons. The decision to cross the nuclear threshold, despite repeated Israeli statements that it would not be the first to introduce nuclear weapons to the Middle East, was reportedly taken in 1967 during the build-up to the 1967 war. Subsequently, Israel has maintained its "non-introduction" pledge while steadily accumulating an arsenal of increasingly sophisticated nuclear weapons and delivery platforms.

The Dimona reactor and associated facilities, located in the Negev Desert, still constitutes the core of the Israeli programme. While Israel's nuclear infrastructure remains shrouded in secrecy, other relevant facilities are believed to include a weapons R&D laboratory at Nahal Soreq, a weapons assembly plant at Yodefat, nuclear weapons storage facilities at Tirosh and Eilabun, and nuclear-capable missile bases at Moshav Zekharya and Sdot Micha (Map 69).

In terms of delivery systems, there is strong evidence that Israel now possesses all three elements of the nuclear triad. Israel's land-based system is based on variants of the nuclear-

Map 69 Israel's known and possible weapons of mass destruction sites

capable Jericho missile. The Jericho-1 is a short-range missile with a range of 500 km, while the two-stage Jericho-2 missile has a range of some 1,500 km. Israel possesses a number of aircraft, such as the F-16I Falcon, that could deliver nuclear weapons, and a large number of cruise missiles that could potentially be launched from land, sea or air. More recently, Israel appears to have developed a sea-based nuclear capability structured around three Dolphin-class diesel submarines supplied by Germany in 2000, and Harpoon missiles provided by the USA that can be adapted to carry a nuclear payload. Among known nuclear powers, Israel is one of only three (alongside the USA and Russia) with credible and reliable land, sea and air-based nuclear delivery systems. Though Israel has never used nuclear weapons, it reportedly deployed weapons in readiness for use during the Yom Kippur war of 1973 and the 1991 Gulf War. For obvious reasons, Israel has not signed the Nuclear Non-proliferation Treaty (NPT) and so is not subject to inspections by the International Atomic Energy Agency (IAEA).

If the Israeli nuclear programme remains shrouded in mystery, then even less is known about Israeli chemical and biological capabilities. However, one of the world's foremost experts on Israel's unconventional weapons programmes, Avner Cohen, has argued that there is little doubt that Israel has developed, stockpiled and, perhaps, deployed chemical weapons at some point since 1947. The centrepiece of Israeli R&D on biological and chemical capabilities is the Israel Institute for Biological Research at Ness Ziona, established in 1952 in an effort to create what Prime Minister Ben-Gurion referred to as a cheap, "non-conventional" weapons capability.

At various times, Israeli leaders have made public statements that strongly imply that Israel possesses offensive chemical weapons capability. In the build-up to the 1991 Gulf War, for example, the Israeli Minister of Science publicly suggested that if Iraq were to use chemical weapons against Israel, Israel should respond "with the same merchandise". Fuelling suspicions has been the Israeli refusal to accede to either of the major international arms control treaties dealing with chemical and biological weapons. Israel signed the 1992 Chemical Weapons Convention (CWC) but has yet to ratify the treaty. Israel remains a non-signatory to the Biological Weapons Convention (BWC).

Iran's nuclear programme

Located centrally in the most volatile region on the globe, Iran has nonetheless avoided direct conflict for the past eighteen years and has pursued its own, rather enclosed, path of development. It is a major power in the Middle East, a keystone of Islam, but a state which remains unpredictable and, from the Western viewpoint, relatively little known. As a result, there is apprehension, particularly in the West, about the possibility that Iran might become a nuclear power, develop other weapons of mass destruction or enhance its missile capability.

On a regional scale, Iran's neighbours include India, Pakistan, China, Russia and Israel, all of which are nuclear powers. On a local scale, it separates Iraq and Afghanistan, the two foci of what has been identified by the USA as the "war on terror". Iran has been singled out by the USA as part of the axis of evil and a rogue state. It is almost completely encircled by major US military bases which are located in Turkey, Iraq, Kuwait, Bahrain, Qatar, Oman, Pakistan, Afghanistan and Kyrgyzstan. Given the extreme pressure under which it has been placed, the leadership in Iran has, since 1989, needed to take serious account of its own defence requirements. It has forged links, overt and covert, with various key powers within the region and has also sought to influence events by asymmetrical activity. In particular, Hezbollah has played a significant role in the Middle East.

While Iran's weaponry in general is seen as a major issue, the key component is the potential for nuclear arms. Pakistan, a state which abuts directly on to Iran, has, as a result of a secretive programme, developed nuclear weapons and become an admitted nuclear non-signatory of the Nuclear NPT. This might well be the approach by the current leadership in Iran. However, Pakistan has been accepted as a cornerstone in the fight against global terrorism and its successful nuclear programme has resulted in virtually no constraints being placed upon it.

Even more secretive has been the nuclear weapons development programme of Israel which, one must assume, has been fully supported throughout by the USA. Thus, two states, one openly hostile and the other at least inimical, are armed with nuclear weapons. It is difficult to see why such pressure is being applied to Iran when little was used against Pakistan and none at all against Israel. Furthermore, it can be argued that the possession of nuclear weapons is defensive rather than offensive, a guarantee against attack. Also, retribution would be swift as the actual use of nuclear weapons would be likely to evoke an immediate response in kind.

During the period of the Cold War, Iran was an important ally of the West and constituted, with Saudi Arabia, the "twin pillar" policy of the USA in the region. With US assistance it was able to establish the Tehran Nuclear Research Center during the 1960s and to recruit a range of nuclear scientists. With a view to conserving Iran's petroleum assets following the 1974 oil crisis but also with thoughts on a nuclear weapons option, the Shah established a programme to develop nuclear power. This was shattered as a result of the 1979 Iranian Revolution when Ayatollah Khomeini, the Supreme Leader of the Islamic Republic, opposed nuclear technology on religious grounds. Therefore, from 1979 to 1989 the programme comprised small-scale research.

However, Khomeini's successor as Supreme Leader, Ayatollah Khamenei, and the new President, Rafsanjani, decided to expand the nuclear programme, overtly in the case of power generation and covertly for fuel cycle activities, for which support was received from Pakistan. The key focus during the 1990s was the Bushehr Nuclear Power Reactor for which assistance was received from Russia. However, the USA was able to constrain further Russian collaboration in the same way that it had enforced the end of cooperation with China in 1997. As a result of these vicissitudes it seems reasonable to conclude that there is strong domestic support for the development of nuclear energy but significant opposition to the development of nuclear weapons. During the new century, Iran has faced the dilemma of whether to risk referral to the UN Security Council which might result in economic sanctions and possibly military action or whether to suspend its various enrichment-related activities. The new government of President Ahmadinejad has adopted a more strident approach as a result of which it has become clear that Iran rejects the permanent cessation of its enrichment and reprocessing programmes.

In contrast, the Islamic Republic has a clear and consistent policy on both chemical and biological weapons to which there is a strong religious objection. It is a signatory of the CWC and the BWC. However, it must be stated that similar arguments concerning religion, the law and national interest have been made with regard to nuclear weapons and the NPT. Any chemical or biological weapons programme appears to be largely shrouded in mystery although it is clear that an interest in the subject was sparked during the Iran–Iraq War. It is possible that small-scale experimentation has developed the potential to produce such weapons should they be required in an emergency.

As the conventional weaponry of Iran was reduced and degraded as a result of the Iran–Iraq War (1980–8) so there was an increasing awareness of the capability of ballistic missiles. By the later stages of the war, Iraq is said to have deployed some 800 Scuds. By the mid-1980s Iran, with help from China, had developed short-range rocketry and by the later 1980s had acquired Scud-B missiles from Libya, Syria and later North Korea. Efforts to enhance its missile arsenal continued following the war but it is thought that the current range is limited to 250 km. The Shahab-3 ballistic missile is said to have been produced but there must remain some doubt about this.

The ambiguity that surrounds Iran's nuclear intentions is inevitable given that many of the technologies, materials and processes needed to produce a nuclear weapon are indistinguishable from those needed for a peaceful nuclear power programme. For example, the same uranium-enrichment facilities can be used to produce low-enriched uranium (LEU) for use as fuel in a nuclear power reactor or highly enriched uranium (HEU) for use in a nuclear weapon. Similarly, as part of its normal function, a nuclear power reactor produces plutonium-237, which can then be extracted and used in a nuclear weapon. Thus, while uranium-enrichment facilities and nuclear power reactors can be used in a nuclear weapons programme, they can also be used for entirely peaceful purposes.

Iran claims its intentions are peaceful and, to date, there is no hard evidence to refute this claim. Iran has a "safeguard" agreement with the IAEA allowing the IAEA to inspect its nuclear facilities, and has even signed a so-called "additional protocol" whereby the IAEA can conduct no-notice inspections of suspect facilities. Hence, the present disposition of Iran's nuclear facilities is relatively well known. Following the acknowledgement of Iran's undeclared nuclear activities from February 2003, the IAEA has carried out detailed investigations to substantiate all aspects of what had been declared. There are still some problem areas but in general a comprehensive picture of the current status has been achieved. Among the core components of Iran's nuclear programme is the Tehran Nuclear Research Center (Map 70), which has a range of production facilities and laboratories and also operates a 5-megawatt nuclear research reactor. The reactor is capable of producing about 600 grams of plutonium per year, meaning that it would take seventeen years to produce sufficient for a nuclear weapon. The facility is closely monitored by the IAEA. The Esfahan Nuclear Technology Center operates a number of small research reactors and a uranium conversion facility. The latter is used to convert yellowcake (a refined form of uranium ore) into uranium hexafluoride (the feeder gas for uranium enrichment). Uranium enrichment takes place at a hardened, underground facility near Natanz. The site was visited in February 2003 by IAEA Director General Mohamed El Baradei who reported that 160 centrifuges were operational. To produce enough HEU for nuclear weapons requires tens of thousands of centrifuges operating for an extended period of time (measured in years). The Natanz facility appears incapable of operating at this level of production for the foreseeable future.

Iran's main nuclear power reactor is located approximately 10 miles south of Bushehr. Work on two power reactors began initially in the mid-1970s, but was interrupted first by the Islamic Revolution, and then by the Iran–Iraq War. Work resumed on one of the reactors in 1995 after Iran signed a deal with a state-owned Russian company to complete construction. To produce bomb-grade material from this reactor would be difficult and would require a sophisticated facility to separate the plutonium from the spent reactor fuel. There is currently no evidence that Iran possesses such a facility. At Arak construction of a nuclear research reactor has begun and a heavy water production plant is partially operational. There are uranium mines and mills at Gchine and Saghand and possibly a pilot uranium laser enrichment plant, which may have been dismantled, at Lashkar Ab'ad. Thus there is a spread of sites from north of Tehran to the Gulf.

From the material collected by the IAEA, it can be assumed that Iran is still several years away from acquiring a nuclear weapons capability. At present, research reactors are too small for viable military production. The one possible option appears to be the use of multiple civilian pilot scale plants but this would run a high risk of detection by the IAEA. A covert programme on any scale would also be problematic for the same reason. The other possibility is that Iran succeeds in procuring fissile material from a third party. This would eliminate the need for uranium enrichment, which is, by some distance, the most time-consuming, costly and technologically demanding stage in the production of nuclear weapons.

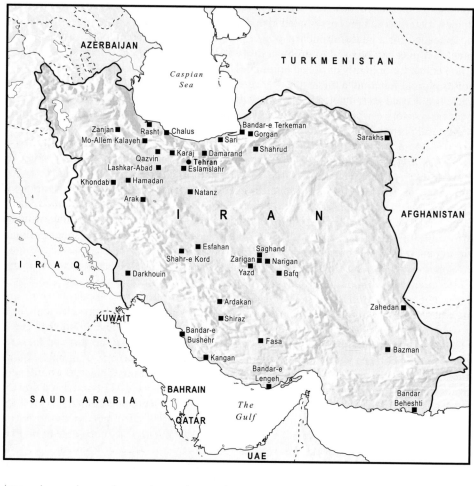

Map 70 Iran's weapons of mass destruction sites

As indicated, the chemical and biological weapons programme, should one exist, is far more difficult to evaluate. It is known that chemical weapons were produced during the Iran–Iraq War but is not known whether production continued after 1988. However, the assumption must be that Iran retains the elements of a chemical weapons programme and facilities are thought to exist at Esfahan, Parchin, Qazvin and Damghan. The biological weapons programme is even more conjectural and only one facility, at Damghan, is normally listed. Two points need to be borne in mind: first, Iran has probably obtained external assistance to enhance whatever programmes it has retained, and second, there may be "dual-use" possibilities. It is not unreasonable to suppose that should the political decision be made, chemical and biological weapons could be obtained.

A further point concerns delivery systems and defence. Since the middle of the Iran–Iraq War, Iran has established what can only be described as the second most technologically advanced ballistic missiles programme in the Middle East but one which is a long way from that of Israel. Already, the ballistic missile capability poses a threat to Iraq, the Gulf countries and neighbouring areas of Central and Southern Asia. The Shahab-3 (No-dong) with a range of 1,300 km would pose threats to Israel, most of Saudi Arabia and Turkey, almost all of Pakistan and a sizeable area of north-western India. Doubt is expressed about the accuracy achieved so far and this would affect their deployment in military terms. However, should the missiles be armed with chemical or biological weapons this would greatly complicate any military operations directed against Iran. Should the Shahab-3 be reconfigured as a nuclear weapon, this would be a major strategic threat to many parts of the region including the US military bases.

The development of more sophisticated ballistic missiles, like that of the other components of the weapons programme, entails potential advantages and disadvantages. In the end, it will be a political decision. At present, missile-related facilities are spread widely throughout Iran although there is a clear concentration around Tehran. Around the borders, there are facilities at Tabriz, Khorramabad, Abadan, Bandar Abbas, Sarji, Mashhad and Shahrud. In addition, many of the nuclear facilities are defended by missile bases, the major concentration being at Parchin.

Iran has, since the demise of the Shah, been considered a threat to US regional and global interests. With the mounting problems in Iraq and Afghanistan, the USA would appear to have limited capacity to attack Iran, other than possibly through air strikes. Nonetheless, US attacks on Afghanistan and Iraq must have caused a certain level of unease in Tehran.

An alternative strategy might be an attack from Israel which has had a nuclear capability since the late 1960s. Since 1979 Israel has regarded Iran as its greatest threat within the region. New F15l and F16l aircraft delivered from the USA have brought the Iranian nuclear facilities well within range and the Israeli air force has also acquired earth-penetrating bombs for use against subterranean facilities. As a result of activities in Iraq, the Israeli Defence Force has developed even closer operational links with the US military, offering the possibility of combined operations under Israeli cover. However, the closeness of the relationship between the two states is well known throughout the Middle East and any Israeli action would implicate the USA.

Any military response from Iran would be limited, since much of its front-line materiel is obsolete. However, asymmetrical warfare would certainly be waged and could greatly influence events in Iraq in particular. Iran would undoubtedly encourage action by Hezbollah and the capability of the Revolutionary Guard should not be underestimated. A major target would probably be the oilfields of the Gulf and the route ways to them including the Strait of Hormuz. Attacks on US bases in Iraq would also be likely, either directly or via proxy militias. This would leave the approximately 130,000 US troops in Iraq in a desperately precarious position. As a result, although militarily Iran is weak compared with the IDF and the USA, it is nonetheless in a very strong position. Furthermore, military action, in particular the use of nuclear weapons, to deter the development of nuclear weapons would reduce the already low standing of the USA in the world in general and throughout the Islamic part of it in particular.

Terrorism in the Middle East

The issue of terrorism in the Middle East inevitably raises difficult definitional issues. Hezbollah, Hamas and Al Qaeda in Iraq are strikingly different in terms of organisation, goals and targeting strategy, yet all are lumped together by the US State Department in the same category of "foreign terrorist groups". The distinction between "terrorism" and "legitimate armed resistance" is very much in the eye of the beholder, which means that reaching an international consensus on the definition of terrorism remains stubbornly elusive. As a tactic, terrorism is generally defined as the deliberate use of violence against civilians to induce terror, and thereby achieve a specific political purpose. This imparts some definitional clarity but also requires one to accept terrorism is not the sole preserve of non-state actors. In so far as states use violence to terrorise the civilians of other states, they too can be guilty of committing acts of terrorism.

With this in mind, it seems clear that the Middle East has suffered more terrorism than any other region over recent decades (Map 71). To many, indeed, the Middle East and terrorism have become so intertwined that it is easy to forget that the emergence of large-scale terrorism in the region is a relatively recent phenomenon. Among the earliest, and most successful, terrorist campaigns in the region were those conducted by various Jewish extremist groups, such as Irgun, under the leadership of future Israeli Prime Minister Menachem Begin. From 1944–7, Irgun launched a campaign of terror designed to drive the British out of Palestine. This included the bombing of the King David Hotel in Jerusalem (killing ninety-one civilians), an attack on the British Officer's Club in Jerusalem that killed seventeen military officers, and multiple assassinations and attacks on infrastructure. Subsequent to the establishment of the state of Israel, violent Jewish extremist groups dissipated, but most of the region's future terrorism would be connected, in one way or another, to Israel's presence in the Middle East. While, today, Islamic Fundamentalism takes the blame for terrorism, the most successful fundamentalists in the Middle East have been the Zionists.

During the 1970s, in response to the failure of Arab states to force a solution to the Palestinian problem on Israel, numerous Palestinian groups emerged to champion the cause of Palestinian nationalism. Confronted with the might of the Israeli Defence Forces, these groups generally opted for terror tactics to advance their cause. Along with several high-profile hijackings, assassinations and kidnappings, Palestinian nationalist groups also staged the "Munich massacre" in 1972, which resulted in the deaths of eleven Israeli Olympic athletes. The group responsible, Black September, was a splinter of the most prominent Palestinian group, Fatah, led by Yasser Arafat. Over time, Arafat succeeded in organising most of the Palestinian groups under the umbrella of the PLO. Though factions of the PLO continued to conduct terrorist operations, against mainly Israeli targets during the 1980s, by 1988 Arafat was willing to sign off a two-state solution and, as part of the Oslo Accords in 1993, the PLO officially recognised Israel's right to exist. Thereafter, the mantle of violent Palestinian resistance passed to Hamas, an organisation founded in 1987, reportedly with Israeli assistance, as a counterbalance to Arafat's secular, nationalist Fatah group.

Throughout the 1980s and 1990s, Hamas was responsible for numerous suicide bombings against Israeli targets. More often than not, the targets were Israeli civilians. By contrast, the Hezbollah movement, although characterised as a terrorist group by many Western governments, has always denied targeting civilians. A predominantly Shi'a movement, Hezbollah was formed in the early 1980s to spearhead resistance against the Israeli invasion of Lebanon. Following the withdrawal of Israeli troops in 2000, and the failure of the Israeli Defence Forces to defeat Hezbollah in the 2006 war, Hezbollah's positive reputation in the Middle East reached unprecedented levels. As a political, social, economic and military movement that has, at times, used terror tactics (such as targeting Israeli cities with rockets in 2006), Hezbollah neatly encapsulates the difficulties involved in defining terrorism. Though classified as a terrorist group by many Western governments, the UK considers only Hezbollah's military wing as a terror group, and throughout much of the rest of the world Hezbollah is considered a legitimate organisational expression of resistance to Israel.

The second major strain of terrorism in the Middle East is of relatively recent vintage. During the 1980s and 1990s, several authoritarian Arab regimes that were viewed as corrupt and devoid of Islamic virtue became targets of a new breed of radical Islamic organisation that used terror as a weapon against the established Middle Eastern order. Egypt, in particular,

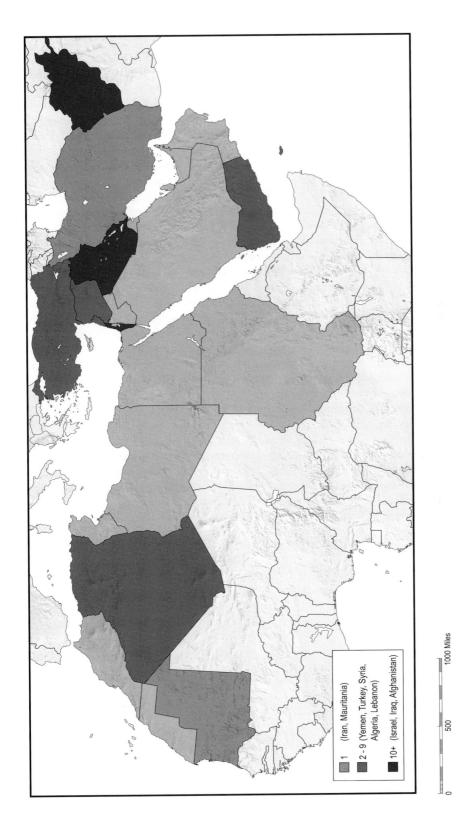

1	(Iran, Mauritania)
2 - 9	(Yemen, Turkey, Syria, Algeria, Lebanon)
10+	(Israel, Iraq, Afghanistan)

0   500   1000 Miles

0   500   1000   1500 Kilometres

Map 71 Terrorist incidents (2008)

suffered greatly from Islamist terror attacks during the 1980s and 1990s. From the early 1990s, the Algerian military's cancellation of elections won by the Islamic Salvation Front resulted in a decade of sustained terrorist assaults against the Algerian military regime. The Al Qaeda network, composed mostly of radical Sunnis, emerged during the 1990s mainly in response to the presence of US troops in Saudi Arabia. After a number of prominent terror attacks against Western targets, the group staged the most spectacular terrorist attack of the modern era on 11 September 2001.

Subsequently, Al Qaeda evolved into a kind of franchise operation, with affiliated groups springing up throughout the world. In the Middle East, the most active of these offshoots have been Al Qaeda in Iraq (AQI) and Al Qaeda in the *Maghrib* (AQM). AQI became a significant force in Iraq following the US invasion of Iraq and the subsequent failure to establish a functioning state. At various stages AQI controlled entire Iraqi provinces, including Anbar and Diyala, as well as large swathes of Baghdad. Known for their elaborate mass-casualty suicide attacks, AQI specifically targeted Iraq's Shi'a civilian population in an attempt to tip the country into full-scale civil war. An AQI attack on the Al-Askari Mosque, which contains the mausoleums of three of the twelve Shi'a imams, in February 2006 came close to achieving this goal. After overplaying its hand and seriously alienating the indigenous Sunni population, AQI became the main target of the "Awakening" movement in Iraq, and is now largely considered a spent force. The Awakening movement (basically groups of former Sunni insurgents organised along tribal lines) has greatly reduced the overall level of insurgent violence in Iraq. At the same time, in 2009, Iraq still remains the most violent country on Earth, with between 400 and 500 civilians dying per month, and suicide attacks taking place on an almost daily basis.

The other main locus of violence in the region is Afghanistan. A resurgent Taliban continues to make inroads against coalition and Afghan forces and now controls anywhere between 50 and 70 per cent of Afghan territory. Each year since 2004 has been more violent than the preceding year, and the Taliban has begun to utilise the same terror tactics, such as mass-casualty suicide bombings, that created such chaos in Iraq. The continued violence in Iraq and Afghanistan, as well as the use of terrorist tactics against Israel, is now driven by a dangerous fusion of nationalism and Islamic fervour. Whether the use of violent terrorist tactics to drive out a foreign occupier is best characterised as "legitimate resistance", "terrorism", or an "insurgency" very much depends on the perspective from which it is viewed.

Political issues

Democracy and freedom in the Middle East

Viewed from one perspective, Map 72, illustrating the levels of political rights and civil liberties in the Middle East, makes for dismal viewing. Only two countries, Cyprus (Greek) and Israel are classified by Freedom House as "Free". A further nine states (Afghanistan, Bahrain, Jordan, Kuwait, Lebanon, Mauritania, Morocco, Turkey and Yemen) fall into the "Partly Free" category, while the plurality of Middle Eastern states are classified as "Not Free". From a more positive perspective, however, almost all states in the region are now freer (by Freedom House's measure) than they were in 1998. Indeed, improvements in five of the states (Afghanistan, Bahrain, Mauritania, Lebanon and Yemen) over this period have been sufficient to lift them out of the "Not Free" category and into the "Partly Free" category. Hence, over the last decade or so, the general trend has been towards greater freedom in the Middle East. There is also quite wide variation within each category. For example, Turkey falls into the partly free category but is numerically close to the free category in both political rights and civil liberties.

It is also reasonable to observe that the measure produced by Freedom House inevitably inspires controversy. The measure is based mainly on the views of Western experts as to the meaning and conceptualisation of the term freedom (and, therefore, democracy), and so this measure is clearly open to accusations of cultural bias. Specifically, freedom is conceptualised in the negative (as the absence of restraint) rather than in the positive (freedom from hunger/poverty, etc). Having said this, the data clearly indicate that, despite a stated commitment on the part of the USA to spread democracy throughout the Middle East, progress has been generally meagre. The only unambiguously free/democratic state is Cyprus. In order to qualify for membership of the European Union, Cyprus was required to meet very stringent standards on all aspects of rights and liberties. Israel meets the criteria for a free state, but only when the Occupied Territories are removed from consideration. Clearly, the populations of Gaza and the West Bank enjoy very little in the way of meaningful political rights and civil liberties.

Explaining the low performance on freedom and democracy indicators of the Middle East as a region is complex. Some have argued that Islam is incompatible with democracy, while others contend that there is something buried deep within Arab culture that prevents democracy from taking root. These are simplistic arguments, however, that ignore the broader context within which Middle Eastern states have existed, particularly since 1945. In several of the states in the region (notably Iraq, Syria and Bahrain) minority groups have traditionally governed over majorities. Though some question the historical significance of the Sunni/Shi'a divide in Iraq, it is a point of historical fact that Sunni Arabs occupied positions of power in the government and armed forces seriously out of proportion to their numerical strength in the population. Apart from a brief interlude (1958–63), Iraq was governed by a Sunni Arab for all of its modern history until 2003. There is an inherent and obvious tension between minority rule and democracy, in which decisions generally reflect the will of the majority.

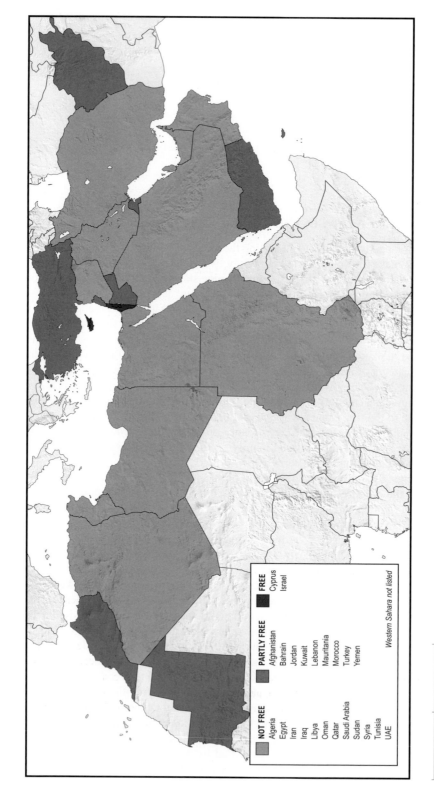

NOT FREE
Algeria
Egypt
Iran
Iraq
Libya
Oman
Qatar
Saudi Arabia
Sudan
Syria
Tunisia
UAE

PARTLY FREE
Afghanistan
Bahrain
Jordan
Kuwait
Lebanon
Mauritania
Morocco
Turkey
Yemen

FREE
Cyprus
Israel

Western Sahara not listed

1000 Miles

500

1000

1500 Kilometres

0

500

0

500

1000

Map 72 Freedom status

In Iraq, democratic elections inverted the traditional power structure, allowing the country's Shi'a population to govern for the first time in its modern history. The violent resistance of the Sunni Arab community to the new democratic order in Iraq was undoubtedly fuelled to a significant extent by the loss of political power and status of the community as a whole.

More generally, the presence of oil in the region has been both blessing and curse. It has enriched elites in those states with ample oil and natural gas resources, but in the process, has also given those elites a strongly vested interest in preserving their status.

Oil and gas have also made the Middle East the key strategic region in the world. Who governs the states of the region has, therefore, mattered to great powers since the 1920s. Generally, the mass popular movements that have emerged in Middle Eastern states: communism, Arab nationalism and Islamism; have all been inimical to Western (and particularly, US) interests. Hence, the likelihood that democracy would bring to power regimes hostile to the West has given the Western bloc, headed by the USA, a powerful incentive to undermine, rather than promote, democracy in the Middle East. Realistically, the West's commitment to the promotion of democracy has been rhetorical rather than genuine. Authoritarian regimes that have served as a bulwark against popular, but threatening, mass movements have traditionally been the regimes of choice for Western powers. There is an inherent tension, therefore, between the West's advocacy of democracy and the sorts of anti-Western, anti-Israeli regimes that would be the likely result of allowing true democracy to take root. A more accurate characterisation of the US position is that democracy is promoted until the people produce the "wrong" result at the ballot box. Hence, elections in Algeria in 1991 that produced a victory for an Islamist party (the Islamic Salvation Front) were annulled in short order by an Algerian military coup. With the strong support of Western powers, including France and the USA, the Algerian military declared a state of emergency and launched a brutal crackdown on the Front's activities. More recently, when Hamas emerged victorious from the 2006 parliamentary elections in the Palestinian territories, its reward for participating in a peaceful democratic process was to be subjected to a stringent economic blockade in its Gaza stronghold.

The lack of freedom and democracy in the Middle East is, therefore, a complex, multifaceted problem. The West generally supports democracy in the region, but only if it produces the "right" result. Unfortunately, the "right" result, the election of moderate, secular, pro-Western and pro-Israeli parties, is also often the least plausible result.

The Kurds

Concentrated on the territories of four Middle Eastern countries (Turkey, Iraq, Iran and Syria), and numbering up to perhaps thirty million people, the Kurds are the Middle East's fourth largest ethnic group (after Arabs, Turks and Persians) and are frequently described as the region's (and perhaps the world's) largest "nation" without a state (Map 73). Divided by the boundaries established in the aftermath of the First World War, the Kurds have been in almost constant rebellion against central authorities of the territories in which they found themselves since the 1920s. More often than not, Kurdish aspirations for autonomy and even outright independence have provoked brutally violent responses from the coercive forces of powerful states, especially in Turkey and Iraq. Trapped by the geopolitical realties of the modern state system of the Middle East that were established by the treaties of Lausanne (1923) and Ankara (1921), the Kurds have never willingly accepted their fate, but are powerless to affect it in the absence of the wholesale reordering of state boundaries in the region.

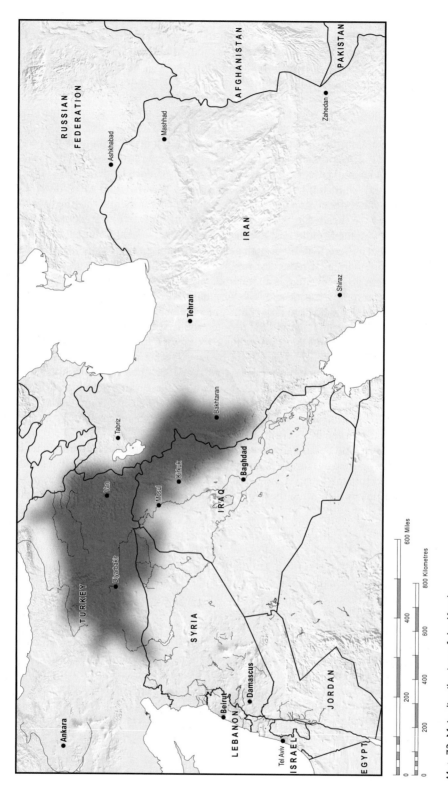

Map 73 Main distribution of the Kurds

The Kurds of Turkey

The largest population of Kurds in the world resides in Turkey. Comprising some fifteen to twenty million people, the Kurds of Turkey constitute about 25 per cent of the country's total population. Though large numbers of Kurds have migrated from the south-east of the country to major urban areas and have been largely assimilated into the Turkish mainstream, the dogged refusal (until recently) of the Turkish state to recognise the existence of distinct Kurdish ethnicity has provoked periodic unrest among the Kurds that remain in the south-eastern portion of the country.

From the late-1970s onwards, this unrest turned violent with the formation of the Marxist-Leninist PKK under the leadership of Abdullah Ocalan. Starting with armed attacks on urban areas, the insurgency had spread to most of the country's Kurdish populated regions by the mid-1980s. The Turkish state responded by placing Kurdish areas under military rule and using widespread, often indiscriminate, violence to destroy the PKK's support bases. The conflict was characterised by numerous atrocities on both sides and had, by the late 1990s cost the lives of between 30,000 and 50,000 people, including many civilians. The capture of Ocalan, the Kurdish leader, in 1999 marked the beginning of the end for the PKK as a serious threat to the Turkish state. The group still conducts sporadic acts of terrorism, striking from its bases in the Qandil Mountains of northern Iraq, actions that have provoked numerous Turkish military incursions into Iraq in recent years. Turkey's ongoing negotiations for entry into the EU have encouraged Ankara to extend cultural and linguistic rights to its Kurdish minority, thereby opening up the prospect of a peaceful resolution to the country's "Kurdish problem".

The Kurds of Iraq

Much like their counterparts in Turkey, Iraq's Kurds have struggled, often violently, against central authority since the 1920s. From 1961 onwards, indeed, Iraqi Kurds conducted an almost permanent low-level civil war against Baghdad until the removal of Saddam Hussein's regime in 2003. Over time, the level of violence the Iraqi government was prepared to use to suppress Kurdish insurrection escalated, culminating in the genocidal Anfal campaign of the late 1980s. Among other things, the Anfal involved the wholesale destruction of thousands of Kurdish towns and villages, the forced relocation of tens of thousands of Kurds and the widespread use of chemical weapons. Following another failed uprising against the Iraqi government in the wake of the 1991 Gulf War, the USA, Britain and France launched Operation Provide Comfort to protect Iraq's Kurdish population from further repression. Shielded by no-fly zones established as part of the operation, the Kurds governed themselves as an autonomous entity until 2003.

Though not without problems, including open hostilities between the two main Iraqi Kurdish political parties, the Kurdistan Democratic Party (KDP) and the Patriotic Union of Kurdistan (PUK), the Kurds' experiment in democratic self-government yielded significant benefits for the Iraqi Kurds. By 2003, the Kurds had established viable political institutions, a vibrant civil society and powerful armed forces. Subsequently, the two main parties maintained a unified front in negotiations over a new constitution for Iraq, and, as a consequence, were able to secure major concessions to sustain their autonomous status in the north. A (mostly) unified Kurdistan Regional Government (KRG) now governs the entire Kurdish territory of northern Iraq, a Kurd (the PUK's Jalal Talabani) is currently Iraq's president, and the Iraqi Kurds, numbering something over four million, have emerged as a powerful force in the political life of the new Iraqi state.

The Kurds of Iran

The second largest population of Kurds in the region inhabit Iran. Approximately eight million Kurds live in Iran, comprising about 12 per cent of the country's total population. Though Iranian Kurdistan was the site of the world's only independent (albeit briefly) Kurdish "state", the Mahabad Republic, established in 1946, the Kurds of Iran have traditionally been less unruly than their counterparts in Iraq. Most Iranian Kurdish political movements have mobilised to secure greater cultural and political freedom rather than to challenge directly the boundaries of the Iranian state, which have endured more or less in their current form for over five hundred years.

The Kurdistan Democratic Party of Iran (KDPI), related to, but distinct from the KDP in Iraq, has historically been at the forefront of the Kurdish nationalist cause in Iran and the movement was instrumental in the overthrow of the Shah's regime in 1979. Hopes for greater autonomy from the incoming Islamic regime were soon dashed and open conflict erupted between the Iranian government and the KDPI in the early 1980s, forcing Kurdish militants to retreat to Iraqi Kurdistan. Subsequently, a series of internal disputes splintered the Iranian Kurdish nationalist movement into multiple factions, often at war with each other. This tendency toward factional infighting and the generally good relationship between Iraqi Kurds and the Iranian government has made life extremely difficult for Iranian Kurdish leaders since the 1980s. Many influential Kurdish leaders have been assassinated by Iranian forces with the aid of Iraqi Kurdish parties. The designation of Iran as a member of the "axis of evil" by the George W. Bush administration and the subsequent invasion of Iraq in 2003 gave hope to Iranian Kurdish parties that Iran might be next in line for regime change. Without intervention by a major military power like the USA, however, the Kurds of Iran remain too fractious and fragmented to present much of a threat to the Iranian regime.

The Kurds of Syria

Numbering between 1.5 and 2 million, the Kurds of Syria comprise approximately 10 per cent of the Syrian population. Because of their relatively small number and their scattered geographical distribution, Syrian Kurds have struggled to organise coherently against the repressive Ba'ath regime. As a result, they have garnered much less international attention than their ethnic cohorts in Iraq and Turkey. Syrian Kurds have suffered serious discrimination within this traditionally Arab nationalist state and are among the poorest of Syria's population. Until 2004, when riots in Kurdish areas led to the deaths of forty Kurds and the incarceration of a further 2,000, their cause remained largely unknown to the outside world. Unlike the Kurds inhabiting other Middle Eastern states, the Syrian Kurds have proved unable to organise into durable political movements, and, as a result, are unlikely to gain significant rights from the current Syrian regime without major external intervention.

The Shi'a in the Middle East

The divide between Sunni and Shi'a sects of Islam dates back to the immediate aftermath of the death of the Prophet Muhammed in 632. The point of contention concerned the line of succession to Muhammed, with the Sunnis favouring the selection of a caliph by notables within the community and the Shi'a advocating the direct appointment of the Prophet's son-in-law, Ali ibn Abi Talib. The term Shi'at Ali (supporters of Ali) was used to refer to the latter group, and was later simplified to al-Shiah. The Shi'a were in the minority from the outset, and Ali's quest to lead the Muslim community was repeatedly thwarted. He finally obtained the position in 656, but was murdered five years later.

The assumption of the caliphate by Yazid in 680 prompted widespread Shi'a discontent and culminated in a planned revolt to be led by Ali's second son, Hussein. On his way from Mecca to Kufa, Hussein and a small bunch of followers were intercepted at Karbala and destroyed by the forces of Yazid. Ever after, Karbala itself and the battle associated with its name have occupied a central place in the Shi'a faith and are synonymous with the suffering of the Shi'a. Far from eliminating Shi'ism, the Battle of Karbala served as a rallying cry for the Shi'a against the tyranny and corruption of the Sunni caliphs, ensuring the survival of the Shi'a throughout the centuries.

In the contemporary Muslim world, the Shi'a are outnumbered approximately nine to one by Sunnis and form the minority in most Arab states in the Middle East. The exceptions are Iraq, where Shi'a comprise approximately 60–65 per cent of the population, and Bahrain, where they make up about 65 per cent of the population. The largest population of Shi'a in the Muslim world inhabits Iran, and Iran's holy city of Qom vies with Iraq's Najaf as the centre of religious learning in the Shi'a world. Other states in the region with significant Shi'a populations include Lebanon (32 per cent), Kuwait (30 per cent), Afghanistan (19 per cent) and Saudi Arabia (15 per cent).

Traditionally in the Arab states, the Shi'a have been viewed with suspicion, particularly in Iraq, where the terms "Shi'a" and "Persian" are often viewed as synonymous by the country's Sunni population. Historically oppressed and persecuted, modern Shi'a religious leaders have generally followed a tradition of "quietism", or conscious non-involvement in secular political affairs. In 1979, however, this tradition was rudely shattered by the Islamic Revolution in Iran. The emergence of Ayatollah Khomeini as unchallenged political leader in Iran brought forth the radical new doctrine of *wilayat al-faqih*, which dictated that only religious scholars were deemed qualified to govern according to *Shari'a* Law. Translated into political practice, this has resulted in a hybrid form of political system for Iran.

Iran has an elected parliament and president (though candidates are carefully vetted), but ultimate power remains in the hands of religious leaders, either through their role on the twelve-member Guardian Council (which approves all legislation), or in the person of the Supreme Leader. The Supreme Leader, currently Ayatollah Ali Khamenei, is the highest ranking religious and political leader in Iran, the ultimate source of legal judgement, and the commander-in-chief of the armed forces. Iran's radical experiment in "rule by the jurisprudent" has not been notably successful in delivering clean, efficient government, and the revolutionary zeal inspired by Khomeini has long since dwindled.

However, Iran's influence in the Middle East has increased dramatically in the aftermath of the wars to remove the Taliban and Saddam Hussein in neighbouring Afghanistan and Iraq. The secular, Sunni Ba'athist regime in Iraq had long been perceived (and used) as a counterbalance against the export of radical Shi'ism from Iran. Likewise, the Taliban are a radical Sunni group that came close to open warfare with Tehran at the end of the 1990s. The removal of both regimes by US-led coalitions rid Iran of two bitter adversaries and left Iran's power effectively unchecked in the Gulf region.

The election in Iraq of a coalition of Shi'a religious parties to govern the country has also greatly enhanced Iran's influence. The dominant power within the coalition is the ISCI, originally a religious militia force created, funded and trained by Iran during the 1980s to fight against Saddam Hussein's regime in the Iran–Iraq War. Iran also maintains close links to other Shi'a political parties/militias in Iraq, and there can be little doubt that Iran is currently the most influential external player in Iraq's internal politics. Iran also has close links to the Hezbollah movement in Lebanon, and Hamas in the West Bank and Gaza, both of which Iran considers legitimate resistance movements against Israeli occupation.

The growing influence of Iran in the Middle East, combined with other developments in the region, the replacement of a secular Sunni regime in Baghdad with a religious Shi'a government, the enhanced prestige of Hezbollah after its 2006 war with Israel, and the murmurings of discontent among Shi'a minorities in several Sunni Arab regimes, have led some to speak of a "Shi'a revival", and others to detect an emerging "Shi'a Crescent" in the region (Map 74). The term "Shi'a Crescent" was first used by King Abdullah of Jordan in December 2004 and seemingly referred to an emerging alliance of forces: Iran, Iraq, Syria, Hezbollah and Hamas; that threatened the established, Sunni-dominated order in the Middle East. Subsequently, the use of the term was heavily criticised by many for its implication that sectarian identity rather than political expediency was the defining feature of these relationships. In fact, Hamas is an overwhelmingly Sunni movement, and Syria has a majority Sunni population. For decades, Syria has been ruled by a political/military elite belonging to the Alawi Sect, an obscure minority sect, that some argue shares similarities with Shi'ism. However, Syria's rulers are Ba'athists, and hence, avowedly secular in orientation. A more meaningful way to characterise the Shi'a crescent, therefore, is perhaps to consider it from the perspective of strategic resources rather than political alliances. This would require drawing an alternative crescent that stretches from Iran, through southern Iraq, Kuwait and Bahrain to eastern Saudi Arabia. The territory under this crescent contains most of the region's, and therefore, the world's, reserves of oil. The implication is, therefore, that the bulk of the world's oil reserves are now under the control of states with either majority Shi'a populations (Iran, Iraq and Bahrain), or significant, but restive, Shi'a minorities (Kuwait and Saudi Arabia). However one chooses to conceptualise the term "Shi'a Crescent", it is clear that the increase in influence exercised by Iran in the region is central to the concerns of surrounding Sunni Arab states.

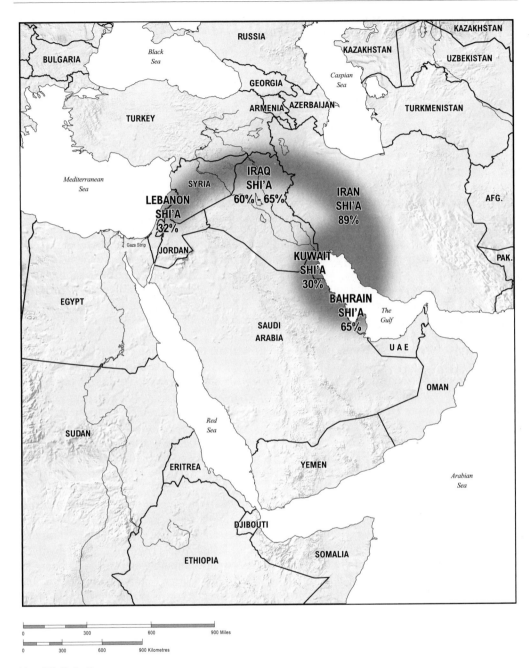

Map 74 Shi'a Crescent

The Israeli–Palestinian question

The 1948 Arab–Israeli War

The 1948 war between the nascent state of Israel and a number of surrounding Arab states was the first of three post-Second World War wars fought between Israel and its Arab neighbours. Of the three, the 1948 war was arguably the most significant for the region. By the war's end, Israel had become a *de facto* state, its territory had expanded by 50 per cent over what the UN's partition plan had envisaged, and somewhere between three-quarters of a million and one million Palestinians had become refugees (Map 75). The region, the Arab world in particular, is still reeling from the effects of what the Israelis refer to as the War of Independence, and most Arabs refer to simply as the Catastrophe.

Like many of the core problems of the Middle East, the genesis of the Arab–Israeli conflict can be traced to the period during and immediately following the First World War. In 1917, the British government declared its "sympathy with Jewish Zionist aspirations" and its favourable attitude towards "the establishment in Palestine of a national home for the Jewish people" in the now infamous Balfour Declaration. The preceding year, the Anglo-French Sykes–Picot Agreement had established spheres of influence for the great powers over former Ottoman territories in the Middle East. Britain's influence/control over Mesopotamia, Transjordan and Palestine was formally codified in the San Remo Agreement and approved by the League of Nations in July 1922. At the time, the population of Palestine was approximately 80 per cent Muslim, 11 per cent Jewish and 9 per cent Christian.

However, over successive decades, mainly in response to increased persecution in Europe, the proportion of Jews in the population increased significantly. Partly in response to this growing influx of Jews into the region, the 1930s witnessed a violent reaction from Palestine's Arab population. The Great Arab Revolt (or Uprising) lasted from 1936 to 1939 and involved riots, strikes and the infliction of violence against the Jewish community and British troops. The British responded by mobilising the Jewish community into armed paramilitary groups that were to provide the backbone of the security and intelligence forces of the new Israeli state. With the aid of these groups, the British successfully suppressed the revolt, often using excessive and indiscriminate violence against the Arab population. Up to 5,000 Arabs were killed during these years, with a further 10,000 wounded. On the Israeli side, the death toll was much lower, perhaps 400–500.

In the aftermath of the Second World War, the UN General Assembly approved Resolution 181 in 1947, which partitioned Palestine into separate Jewish and Arab states. The plan, illustrated in Map 75, allocated some 55 per cent of Palestine's territory for the Jewish state, and 43 per cent for the Arab state. Jerusalem, meanwhile, was to remain under international (UN) administration. This division of territory was clearly unfavourable to the Arabs. With roughly two-thirds of the population, they were allocated significantly less territory than the less numer-

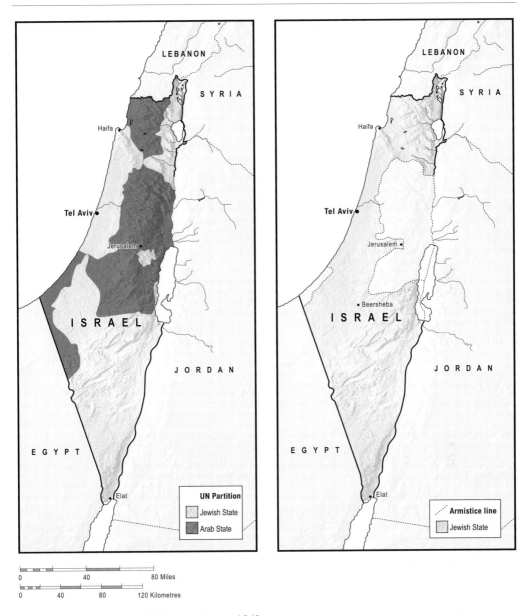

Map 75 Jewish state: pre-1948 war and post-1948 war

ous Jewish population. The UN's approval of the plan was welcomed by the Jewish side, but there was little in the plan to tempt the Arabs.

From the end of December 1947 onwards, Palestine witnessed a serious upsurge in violent incidents. During 1948, the violence began to assume a more organised form. Various Arab armies, including the Arab League's Arab Liberation Army and the Army of the Holy War, intervened during this period, which came to be referred to as the Palestinian Civil War. As the violence intensified in the run-up to the termination of Britain's mandate over Palestine on 15

May 1948, it became clear that Israel would face attack from several of its Arab neighbours. On 14 May, future Israeli Prime Minister David Ben-Gurion declared Israel's independence, which was recognised almost immediately by the USA and the Soviet Union. The following day, the Arab League issued its own statement declaring the UN plan invalid and proclaiming its intention to create a "United State of Palestine" in place of the two-state solution outlined in the UN partition plan. The same day, the military forces of five Arab states: Egypt, Syria, Jordan, Lebanon and Iraq; invaded Israel.

From the outset, Arab forces struggled to compete militarily against the highly motivated Israeli forces. Even with five states participating, the Arabs were heavily outnumbered. By the end of 1948, Israeli forces in the field numbered over one hundred thousand, compared with fewer than half that number on the Arab side. Arab forces were also badly trained, poorly equipped and desperately lacking in coordination. The leaders of the participant states had diverse goals with respect to the war. King Abdullah of Jordan, for example, viewed the war primarily as an opportunity to annex parts of Palestine to Jordan, so the most effective Arab army, the Jordanian-led Arab Legion, focused most of its energies on establishing control over Jerusalem. Elsewhere, the other Arab armies made little progress during the first phase of the war (May–June 1948).

A 28-day UN-brokered truce that expired on 8 July was followed by a series of successful Israeli offensives that resulted in the capture of the strategic cities of Lod and Ramla, and the whole of lower Galilee. Following the breakdown of another truce in October 1948, the final phase of the war saw Israel on the offensive permanently. In October, the Israeli's launched Operation Hiram, which captured all of Upper Galilee, and drove the Lebanese army out of Israel. Operation Horev (December 1948) drove the Egyptian army out of the Negev Desert and led to a ceasefire agreement between the two countries with the Egyptian army surrounded in the Gaza Strip. Between February and July 1949, Israel signed separate armistice agreements with four of the five Arab armies (Iraq was the exception).

As a result of the war, Israel increased its territory significantly over that envisioned by the UN's partition plan. As illustrated by Map 75, post-war Israel was about half as large again as the UN proposal. Of Mandate Palestine, only the West Bank area and the Gaza Strip remained under Arab control. The former would remain under Jordanian control, and the latter under Egyptian control until the next Arab–Israeli war of 1967. The 1948 war also resulted in what remains to this day among the most contentious and emotive issues in the Israeli–Palestinian dispute. Estimates of the number of Palestinians who fled their homes or were forcibly expelled by Israeli forces vary, but usually fall within the 700,000 and 800,000 range. The standard Israeli position is that these Palestinians fled as a result of the war, or because they were encouraged to do so by Arab leaders. The Arabs maintain that there was a deliberate plan (code-named Plan D) on the part of Israeli forces to drive Palestinians off their land in an effort to expand the boundaries of Israel. Either way, there were up to 3.5–4 million Palestinian refugees living in refugee camps throughout the Middle East, often in extremely harsh conditions. This tragic legacy of the 1948 war continues to thwart efforts to find a durable and peaceful solution to the Israeli–Palestinian conflict.

The Palestinian refugee problem

The problem of Palestinian refugees stems mainly from the period immediately prior to, and during, the 1948–9 Arab–Israeli War. In 2009, approximately one-third of the estimated 3.5–4 million refugees still live in refugee camps established in the aftermath of the conflict, and their fate remains one of the key obstacles to an Israeli–Palestinian peace deal (Map 76). The Pales-

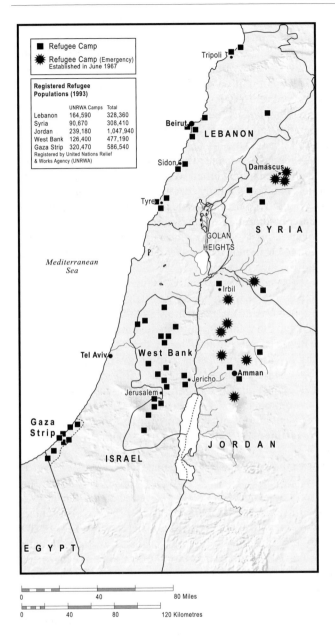

Map 76 Palestinian refugees

tinians demand that all refugees (and their offspring) displaced during the course of the 1948 war have the right to return to their place of origin. From the Israeli perspective, this demand is impossible to meet because the return of so many Palestinians to Israel would in all probability make Israel a majority Arab country. A yawning chasm exists between the two sides on this issue, and in the meanwhile, the Palestinian refugee problem drags on into its seventh decade with no solution in sight.

Estimates of the number of Palestinian refugees vary considerably, though most fall within the 3.5 to 4 million range. The lion's share of these, some 1.8 million, live in Jordan. Unlike other Arab countries, the Jordanians have made citizenship relatively easily available to refugees, and the majority of Palestinians have become integrated into the mainstream of Jordanian economic and social life. Of the total number, only approximately 10–15 per cent still live in thirteen refugee camps located in Jordan. Jordan suffered another influx of Palestinian refugees as the result of the 1967 war, though these are officially classified as "displaced persons" rather than refugees. The situation for refugees in Lebanon is significantly worse. The tenuous ethno-sectarian balance in Lebanon means that only about 25 per cent of the approximately 400,000 refugees in Lebanon have been given citizenship. The rest live in generally squalid conditions in twelve refugee camps dotted in or around the cities of Sidon, Tyre, Beirut and Tripoli. With an average estimated family income of below $400 per month, no prospect of citizenship, no access to state-sponsored health care or education, and few employment possibilities, Lebanon's Palestinian refugee population is among the worst off of all such communities in the region.

Syria's refugee population is estimated to be similar in size to Lebanon's. Of these 400,000–450,000 refugees, approximately one-third reside in twelve refugee camps. Though the Syrian government also denies citizenship to its Palestinian population, refugees do enjoy rights to employment and education and the government provides funding to help maintain the camps. Elsewhere in the Middle East, Saudi Arabia is host to approximately 240,000 refugees, Egypt to 70,000 and various other Gulf States to about 100,000. Kuwait expelled most of its 400,000 or so Palestinian population following the PLO's support of Iraq during the 1990 Iraqi invasion of Kuwait, while Iraq's Palestinian population is dwindling fast as a result of targeted persecution. In addition to refugee populations of neighbouring countries, there are still significant numbers of Palestinians living in camps in the West Bank and the Gaza Strip. In Gaza, for example, there are an estimated 1 million refugees, constituting some 71 per cent of Gaza's total Palestinian population. Nearly half of these live in dire conditions in eight densely populated camps. Of the West Bank's population of 2.5 million, an estimated 700,000 are refugees, one-quarter of whom live in nineteen camps.

The fate of Palestinian refugees is one of the key issues that will need to be addressed in order for a meaningful Israeli–Palestinian/Arab peace deal to take hold, but it is not at all clear how this can be done in a mutually acceptable way. As early as 1949, the Israeli government stated that any solution to the refugee problem must involve refugees being absorbed by surrounding Arab states rather than their return to Israel, and it is highly unlikely that the Israelis will ever accept anything more than a token, symbolic number of returnees for the simple reason that to accept an unlimited right of return would risk creating an Arab–Muslim majority in what is defined as a Jewish state. Israel's fragile demography, therefore, prevents much in the way of compromise on the Israeli side.

On the Arab side, the issue has emotive force because of the manner in which the Palestinian's were originally displaced. Unlike the Israeli version of events, in which Arabs fled to avoid the fighting or because they were ordered to by Arab leaders, Arabs contend that Israel used the 1948 war as an excuse to execute a deliberate and systematic programme of ethnic expulsion. The contents of Plan Daleth (otherwise known as Plan D) are often cited by Arabs as evidence of the intentional nature of the displacement of Arabs, though the plan itself remains open to a variety of interpretations. While the plan's primary intent was "to gain control of the areas of the Hebrew state and defend its borders", it also called for offensive operations against "enemy population centres" that involved "destruction of villages (setting fire to, blowing up, and planting mines in the debris)". That a significant number of Arabs left involuntarily is beyond doubt, but that this reflected a deliberate policy of ethnic cleansing on the part of Israel is less than clear. Regardless of the truth of the matter, Arab states have done little to help the plight of their Palestinian brethren. For example, the Arab League instructed member states not to grant citizenship to Palestinian refugees, an instruction with which all states but Jordan were only too happy to comply. While the intent underlying the instruction was clear, in that to have granted citizenship to Palestinians would have negated the right to return, the effects have been extremely detrimental to the Palestinian refugees themselves. Throughout the Middle East, Palestinian refugees live in poverty, and in legal limbo. Since 1948, three generations of refugees have grown up as stateless individuals in someone else's state. However, the sad reality is that it is probably in nobody's interests, except the Palestinians themselves, to search meaningfully for a permanent solution to the refugee problem.

The 1967 war

The second, and most geopolitically significant of the three major post-Second World War Arab–Israeli wars took place in June 1967 and lasted only six days. The war pitted Israel against the armies of Egypt, Syria and Jordan, together with a small Iraqi force and some negligible contributions from other Arab countries such as Algeria and Kuwait. As with all contentious Arab–Israeli issues, responsibility for initiating the war is hotly disputed. In May 1967, Egyptian President Nasser ordered the United Nations Emergency Force (UNEF) to withdraw from the Sinai where it had been stationed since the end of the Suez Crisis. The Egyptians subsequently closed the Straits of Tiran to Israeli shipping and moved some 100,000 troops and up to 1,000 tanks into the Sinai. On 30 May, Egypt signed a mutual defence treaty with Jordan, having previously concluded a similar agreement with Syria in 1966.

Against a background of escalating border violence between Israel and both Syria and Jordan, Israel launched a pre-emptive strike against the Egyptian air force on 5 June. Though Israel claimed it feared an imminent Egyptian attack, Egypt's forces were arrayed in defensive positions, a fact that is generally acknowledged by all sides. Consequently, Israel's attack might better be defined as "preventive" rather than "pre-emptive". Nonetheless, Israel's surprise attack had devastating consequences for the Arab side. On the first day of the war, Israel succeeded in destroying 300 of Egypt's 450 combat aircraft. The following day, the Israeli air force launched similar raids against the much smaller Jordanian and Syrian air forces and by the end of day two, the Israelis had destroyed over 400 Arab aircraft at a loss of only twenty-six Israeli planes. Subsequently, Israel's air superiority was to prove decisive in the rout of Arab forces. On 7 June, Israel launched multiple attacks against Egyptian forces in the Sinai and, following the fall of the strategic town of Abu-Ageila, the Egyptian Minister of Defence ordered a full-scale retreat of his forces. By the end of 8 June, Israel had effectively completed the capture of the Sinai Peninsula.

Meanwhile on 5 June, in the mistaken belief that Egyptian forces were making progress to the south, King Hussein of Jordan launched attacks into the Jerusalem area. After heavy fighting on 7 June, Israeli troops advanced deep into the West Bank while simultaneously Israeli paratroopers captured the Old City of Jerusalem. In rapid succession, Israeli forces then captured Judea, Hebron and Nablus. To the north, the Syrian front opened up on 5 June with the destruction of two-thirds of the Syrian air force. Subsequently, Israeli forces captured the strategic Golan Heights after meeting cursory resistance from the Syrian forces stationed there.

By 11 June, therefore, Israel had comprehensively defeated all three of its major Arab adversaries and had captured territory that was of huge strategic and political significance. Alongside the Sinai Peninsula (returned to Egypt as part of the 1978 Peace Agreement), Israeli troops now occupied the Gaza Strip, all of the West Bank (including East Jerusalem) and the Golan Heights

(Map 77). These newly captured territories increased the size of Israeli territory by a factor of three and placed over one million Arabs directly under the control of Israel. The Six Day War was a devastating defeat for Arab forces. Against 1,000 Israeli battle deaths, the Egyptians lost 10,000, Jordan over 700 and the Syrians some 2,000. The speed and ease of the Israeli victory was deeply humiliating to the pride of the Arab World. More importantly, the Israeli conquests during the war remain at the heart of the ongoing Israeli–Palestinian conflict.

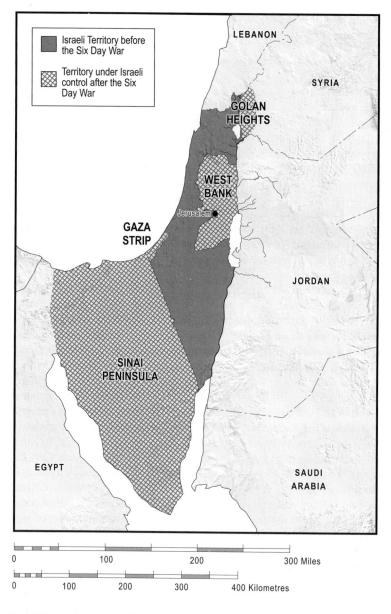

Map 77 Israel after the 1967 war

In November 1967, the UN Security Council adopted Resolution 242 which required the Israelis to withdraw from "territories occupied" during the war in return for a guarantee of peace for all states in the region "within secure and recognized boundaries". The English version of Resolution 242 refers simply to "territories occupied", while the Russian, French and Spanish translations include the definite article prior to the word "territories". Hence, there is ambiguity surrounding how much of the occupied territories Resolution 242 requires the Israelis to withdraw from. Almost immediately, however, the Israelis embarked on a programme of intense settlement construction in Gaza and the West Bank in an attempt to create facts on the ground. Additionally, the Israelis have greatly expanded their presence in Jerusalem with a view to eventually proclaiming the city as the undivided capital of Israel. Most experts consider these efforts by Israel to be in violation of international law.

Of the territories captured by Israel in 1967, the Sinai was returned to Egypt in 1978, and Israel formally dismantled settlements and withdrew from the Gaza Strip in 2005. By 2008, there appears to be some movement towards a negotiated solution between Syria and Israel over the Golan Heights. However, the continued Israeli occupation of the West Bank, and, indeed, the ongoing construction of settlements there generates intense hostility in the region and thwarts the good faith efforts to find a peaceful solution of the Israeli–Palestinian conflict. Thus the fallout from a war fought more than forty years ago continues to poison the atmosphere in the region to the present day.

The Israeli–Palestinian peace process

The first serious steps toward negotiating a peace deal between Israel and the Palestinians took place in the immediate aftermath of the collapse of the Soviet Union and the defeat of Iraq in the Gulf War of 1991. The Soviet Union had consistently been among the strongest great power supporters of the Palestinian cause and its collapse left the Palestinians with no reliable great power protection. Moreover, the PLO, under the leadership of Yasser Arafat, had adopted a strongly pro-Iraqi stance during the Gulf War, and Iraq's devastating defeat left the PLO with few supporters and, therefore, financial backers among Arab states in the region. From the Israeli perspective, the emergence of Israel's major international patron, the USA, as the single superpower in the international system, and the general war-weariness engendered by five years of the Palestinian *Intifada*, led many in Israel to favour a peaceful resolution to their interminable conflict with the Palestinians.

Under these favourable conditions, serious negotiations between the two sides began in 1992–3 in Madrid, culminating in an agreement signed in Oslo in August 1993. What came to be known as the Oslo Accords were presented to the world in an elaborate signing ceremony in Washington DC (Map 78). The negotiations surrounding the Accords were the first time that Israel had consented to deal face to face with Arafat and the PLO. They also marked a key turning point in the history of relations between the two sides because they required the PLO both to recognise the right of Israel to exist and to renounce violence, while on the other side, the Israelis officially recognised the PLO as the competent Palestinian authority. The basic principle underlying the Accords was a "land for peace" deal along the lines originally outlined in UN Security Council Resolution 242 of 1967, by which the Israelis would withdraw incrementally from the Gaza Strip and parts of the West Bank. Vacated areas would be turned over to a newly created Palestinian Authority (effectively the PLO) to govern, in return for which the Palestinian side would have to police these territories effectively to guarantee Israel's security.

Beyond this basic principle, the details were more complex. The Accords divided up the West Bank and Gaza into three zones. Territories designated "Area A" were those under full Palestin-

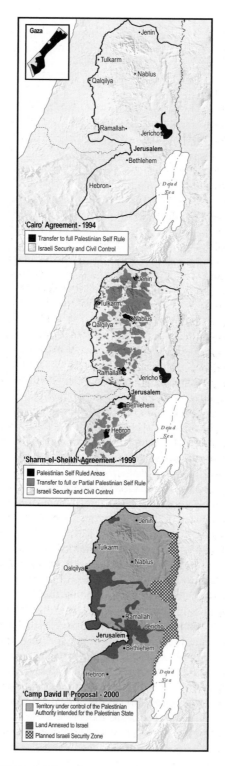

Map 78 Palestinian–Israeli peace agreements

ian control; "Area B" zones were subject to Palestinian civil control, but Israeli security control; zones classified as "Area C" would remain under full Israeli control. Over time, the intention was to change incrementally the designation of territory such that, by the end of the process, the Palestinians would be left with a certain percentage of the West Bank and Gaza classified as Area A, which would then, presumably, have become the Palestinian state. However, Oslo was deliberately vague on any final settlement. Core issues such as the final status of Jerusalem, Israeli settlements, the refugee problem and the physical dimensions of a Palestinian state were left for future negotiations. Negotiations over these and other final settlement issues were to begin no later than 1996 and be completed within five years of Oslo.

Initially, progress on implementing the Accords was slow but steady. The 1994 Cairo Agreement committed Israeli forces to withdraw from Gaza and Jericho, and turned over full control to the Palestinian Authority (Map 78). Then in 1995, the Interim Agreement on the West Bank and the Gaza Strip (Oslo II) gave self-rule to Palestinians in several cities, including parts of Hebron, Jenin, Nablus, Ramallah and Bethlehem, as well as hundreds of villages. However, the process effectively ground to a halt in the aftermath of the assassination in 1995 of Israeli Prime Minister Yitzhak Rabin by a Jewish extremist opposed to Oslo. Rabin had been one of the architects of Oslo and among the strongest supporters of a negotiated solution to the Palestinian issue. Subsequently, a campaign of suicide bombings conducted by radical Palestinian groups opposed to the process (Hamas and Palestinian Islamic Jihad) in the run up to the 1996 Israeli elections solidified support behind the candidacy of Benjamin Netanyahu for Prime Minister. A hard-line member of the Likud Party, Netanyahu was strongly critical of the Oslo process and generally reluctant to fulfil any of Israel's commitments under the Accords.

With considerable American pressure, Netanyahu concluded the Hebron Agreement in 1997, which required Israel to withdraw from 80 per cent of Hebron and undertake a phased withdrawal of Israeli forces from other parts of the West Bank. The 1998 Wye River Memorandum committed Israel to further redeployments and would have turned over approximately 13 per cent of the West Bank to Palestinian control. The memorandum also called for a resumption of final status talks. After alienating the left in Israel by dragging his feet over the peace process, and the right by agreeing to anything, Netanyahu was defeated by Ehud Barak in the 1999 election for Prime Minister. Almost immediately, Barak concluded the Sharm el-Sheikh Memorandum with Arafat (Map 78), which envisaged an Israeli withdrawal from a further 11 per cent of the West Bank and agreed the release of some 300 Palestinian prisoners. The memorandum also called for an intensification of efforts to reach a final status agreement.

As it became clear during the course of 1999 that the Oslo process was running out of steam, behind the scenes negotiations began at Camp David in July 2000 to reach a deal that would encompass all dimensions of the Israeli–Palestinian issue, including such contentious matters as the status of Jerusalem and the return of Palestinian refugees. The hope was that a grand bargain could be struck while President Clinton was still in office. Predictably, accounts of the negotiations vary depending on perspective. According to the Israelis (and many US officials), Arafat rejected the most generous offer the Palestinians had ever received, which would have created a Palestinian state on all of the Gaza Strip and about 91 per cent of the West Bank (after an interim period). The Palestinians claim the Israelis never actually put a clear proposal on the table, and that Israeli "solutions" to the status of Jerusalem and refugees were inadequate. Map 78 illustrates what Palestinian negotiators claim to be the Israeli offer for a Palestinian state. In effect, the West Bank part of the state, although comprising over 90 per cent of the territory, would have been divided into three discrete cantons that were neither connected to each other nor to the Gaza Strip. The Israelis claim their offer would have connected the north and central cantons, but have never released an official map to support this claim.

The outbreak of the Second *Intifada* in September 2000 spelt the death knell of hopes for a comprehensive peace deal. Although President Clinton continued to press for a deal in the dwindling days of his administration, and the two sides met again in January 2001 at Taba and came close to reaching a final status agreement. However, with Clinton out of office, and Barak facing an unsuccessful re-election bid against the hawkish Ariel Sharon, the time for a meaningful peace deal had expired.

Israeli settlements and checkpoints

The construction and ongoing expansion of Israeli settlements on territories captured during the Six Day War is a source of deep resentment in the region and a major obstacle to an eventual peace deal between the Israelis and Palestinians. As of 2009, there are approximately 480,000 Israeli citizens inhabiting settlements constructed on the West Bank, the Golan Heights and in East Jerusalem (Map 79). Settlements on other territories captured by the Israelis during the 1967 war have, over time, been dismantled and the settlers removed. Hence, as a result of the 1979 Egypt–Israel Peace Treaty, Israel agreed to dismantle and remove all settlements from the Sinai Peninsula. This process was completed in 1982. More recently, the then-Prime Minister Ariel Sharon initiated the unilateral disengagement plan in 2005 that involved dismantling all twenty-one settlements in the Gaza Strip and the forcible removal of those settlers unwilling to leave.

The removal of Israeli settlements from the Sinai and Gaza left approximately 480,000 settlers in settlements distributed across the West Bank, the Golan Heights and East Jerusalem. Of this number, some 280,000 live in West Bank settlements. The scale and sophistication of West Bank settlements varies enormously. Some, such as Ariel, Beitar Llit and Modiinilit have stable (and growing) populations that number in the tens of thousands, and possess the permanent infrastructures characteristic of small cities. Other settlements, such as Rotem and Tomer in the Jordan Valley are little more than villages, with populations ranging from the low hundreds to the low tens. Approximately 18,000 Israelis live in settlements on the Golan Heights, captured from Syria during the 1967 war, and a further 180,000 inhabit East Jerusalem, taken from Jordan during the same war.

Disputes over the legal status of Israeli settlements are complex. Overwhelmingly, the international community considers the settlements illegal under international law, specifically, on grounds that they violate Article 46 of the Hague Convention, and Article 49 of the Fourth Geneva Protocol. This widespread sentiment is reflected in several UN Security Council Resolutions (446, 452, 465 and 471) that refer to the settlements as having "no legal validity". Israel has disputed this characterisation, arguing that the Fourth Geneva Convention does not apply because the West Bank was not part of a sovereign state at the time construction of settlements commenced. Moreover, the relevant Security Council Resolutions were all passed under Chapter VI of the UN Charter and are, therefore, non-binding, according to Israel. Israel's position on the legality of settlements is, however, not shared by many in the international community.

The existence of Israeli settlements on the West Bank and East Jerusalem is a source of great resentment to the Palestinians living there. Though the land area occupied by settlements is less than 3 per cent of the total area of the West Bank, the infrastructure associated with the settlements takes up nearly 40 per cent of the total area. This infrastructure includes a network of roads designed to connect settlements to each other and to Israel. Palestinians are excluded from using many of these roads and, for security reasons, these roads are flanked by buffer zones on either side. Further complicating life for the Palestinians is the ubiquitous presence of checkpoints throughout the West Bank. In 2006, there were estimated to be over 500 physical obsta-

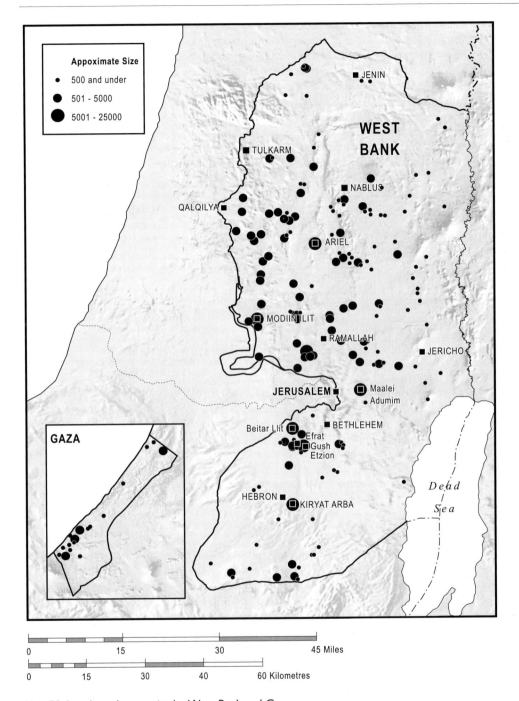

Map 79 Israeli settlements in the West Bank and Gaza

cles to travel for Palestinians in the West Bank, including eighty checkpoints manned by the Israeli military. The exact number of checkpoints in place at any given time varies considerably because many are temporary blocks aimed at countering an alleged specific terrorist threat. The number of checkpoints in operation escalated rapidly with the onset of the Second *Intifada* in 2000, but there is evidence that the construction of the separation wall has led to a decline in numbers over recent years. Overall, this network of roads and checkpoints results in the fragmentation of Palestinian territory in the West Bank into three cantons: north, centre and south; and greatly increases the difficulty experienced in travelling between the three areas.

Though none of the various peace processes launched by the Israelis and Palestinians have required Israel to cease settlement construction, still less to dismantle existing settlements, the ongoing construction of new settlements, most notably in East Jerusalem, and the expansion of existing settlements has been a constant bone of contention between the two sides. Even President George W. Bush described ongoing settlement activity as "unhelpful" to efforts to craft a peace deal in the waning days of his administration. For the Palestinians, the Israeli policy is geared towards grabbing land and thereby creating facts on the ground that will reduce the territory of any future Palestinian state. While "unauthorised" settlements will have to be removed by the Israelis as part of any final status agreement, it is highly unlikely that major settlements, such as Ariel, will ever be dismantled. Any future peace deal, therefore, is likely to require a land swap whereby those major Israeli settlements that fall on the Palestinian side of the green line will become officially part of Israel, in return for which the Palestinians will gain land that currently falls on the Israeli side.

More problematic for the Israeli government will be those settlements that have been "authorised" by the government but which have populations in the hundreds, or low thousands. Often these smaller settlements are populated by extremists who believe strongly that much of the West Bank is part of the original Jewish homeland and are unlikely to leave without violent resistance. Leaving these settlements in place, however, will require the Israeli government to provide for their security and to guarantee transportation links to and from Israel proper. This will result in an "independent" Palestinian state, much of which is still controlled by Israel. This is an untenable solution. Ultimately, therefore, the probability of a future two-state solution may depend less on the Palestinians and more on the willingness of the Israeli government to confront and evict its own citizens inhabiting settlements on the West Bank.

The separation barrier

Almost every aspect of the Israeli barrier that separates Israelis from Palestinians roughly along the 1949 armistice line (the Green Line) is mired in controversy (Map 80). The Israelis generally refer to the barrier as the "security fence" while opponents prefer terms like "apartheid wall". The vast majority of the barrier is fencing, protected by trenches and exclusion areas. About 10 per cent of the barrier, located mostly in urban areas, comprises concrete walls up to a height of 8 m. Though construction is not yet complete, the barrier will eventually stretch for nearly 450 miles when it is finished in 2010.

The original idea for some form of physical separation between Israel and Palestinian-populated territory emerged in response to violent infiltrations into Israel from the Gaza Strip in the early 1990s. Responding to a series of attacks, the then-Israeli Prime Minister Yitzhak Rabin ordered construction of a physical barrier surrounding the Gaza Strip in 1994, justifying the move as an effort to protect Israelis from terrorism. The same justification was advanced in the wake of the serious violence associated with the Second or al-Aqsa *Intifada* that erupted in the West Bank in 2000. After experiencing a wave of suicide bombings in Israeli cities, the government of Ariel Sharon authorised the construction of the West Bank barrier in 2002.

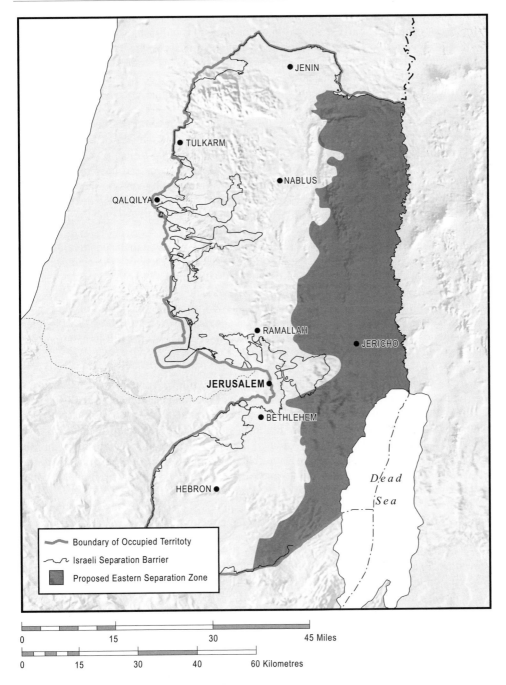

Map 80 Occupied West Bank: separation barrier

The trajectory of the barrier follows the Green Line, but only very approximately. In fact, only about 20 per cent of the barrier is actually on the Green Line. For most of the remainder of its path, the barrier is built on West Bank territory, and in some areas, the barrier cuts deep into Palestinian territory to ensure that major Israeli settlements, such as Ariel and Gush Etzion fall on the Israeli side of the barrier. Consequently, the barrier follows a complicated course that in some cases divides Palestinian settlements and in others almost completely encircles Palestinian villages.

The Israeli claim that the barrier has had a significant positive impact on the security of Israeli citizens appears to be supported by most of the available data. The number of Israeli civilians killed by Palestinian infiltrators declined from its high point in 2002 to just over 100 in 2003, and to ten in 2004. How much of this reduction is due to the barrier and how much to other factors, such as the truce between Israelis and Palestinians, is difficult to assess. There may also be some benefits for the Palestinians. For example, the barrier has significantly reduced the need for Israeli military incursions into the West Bank, and the number of Israeli checkpoints on the West Bank has almost halved since work began on the barrier.

However, the negative effects on the Palestinian people clearly outweigh the positive. Notably, work on the barrier has required the Israelis to confiscate large tracts of Palestinian land, to demolish Palestinian settlements and to destroy up to 100,000 citrus and olive trees. According to the UN, the territory between the barrier and the Green Line is home to nearly 50,000 Palestinians who are now effectively cut off from their ethnic kin on the other side of the barrier. Likewise, the barrier's route through Jerusalem and its environs will cut off over 50,000 Palestinians from the rest of the city. On the Israeli side, nearly 60,000 citizens still live on the West Bank to the east of the barrier. Their future fate is unclear.

Legal rulings on the status of the barrier tend to confirm the arguments of its opponents. The Israeli Supreme Court has twice stepped in to order the Israeli government to change the route of the barrier in order to minimise the negative effects on Palestinians, but has generally held the construction of the barrier to be a reasonable and legal response to the threat of terrorism against the Israeli population. International legal opinion has been sharply critical of the barrier. In 2003, the USA had to veto a UN Security Council Resolution calling for the destruction of the barrier where it deviated from the Green Line, but the same resolution was adopted by the General Assembly with only four dissenting votes. In 2004, the ICJ issued an opinion that called for the barrier to be removed in its entirety as "contrary to international law".

The construction of the barrier is almost universally opposed by the international community, and even the USA has, at times, been critical of the negative impact of the barrier on prospects for an eventual peace between Israel and the Palestinians. The major fear on the Palestinian side is that the Israelis are in the process of unilaterally creating a *de facto* boundary between Israel and an eventual Palestinian state that includes significant tracts of Palestinian land and that, when taken in conjunction with the Israeli military's "exclusion zones" along the entirety of the Jordan River, will effectively seal off the West Bank from the outside world. It is difficult to see how this can lead to a mutually acceptable peace deal between Israelis and Palestinians.

Iraq

Ethnic distribution

The last reliable census that broke down Iraq's population into ethnic groups took place in 1957. Subsequently, the Ba'ath Party regime systematically manipulated census data to increase the proportion of Arabs at the expense of other ethnicities. As a result, attaching numbers to Iraq's large number of ethnic and sectarian communities will be a matter of educated speculation until an up-to-date census is conducted. Ethnically, Iraq is approximately 75–80 per cent Arab, 15–20 per cent Kurd, with the remainder a mix of Assyrians, Turkomans, Chaldeans and various other smaller groups (Map 81). The exact number of Turkomans is a source of controversy. Turkoman political leaders claim that there are more than three million Turkomans in Iraq, which would put them at about 10 per cent of the total population. The available evidence does not support this high estimate, however. Opinion poll surveys conducted in Iraq using random sampling techniques suggest that Turkomans comprise no more than 2–3 per cent of the population, and support for ethnic Turkoman political parties in recent elections implies that Turkoman numbers may be as low as 300,000–500,000. Iraq's Assyrian and Chaldean populations, both primarily Christian by religion, may have numbered up to one million until recently. However, the Christian population, mainly located in volatile northern areas such as Mosul and Kirkuk, has been viciously targeted by insurgent groups and the number of Christians remaining in Iraq may soon diminish to insignificant levels.

Breaking down Iraq's population by religious sect is even more controversial and challenging, because the category "sect" has never been used in Iraqi censuses, and moreover, there has been a degree of intermarriage between Sunni and Shi'a over the years. On the basis of electoral support for obviously sectarian parties and coalitions, a rough but reasonable estimate of Shi'a numerical strength would be over 60 per cent of the total population. Though Shi'a (Fayli) Kurds exist, particularly along the border with Iran, in Baghdad and in the Sinjar district of Nineveh governorate, the majority of Kurds are Sunnis. Together with the significant population of Sunni Arabs, the total Sunni share of the population is most likely somewhere under 40 per cent (Map 81).

In terms of geographical distribution, the Kurds are the most concentrated. Though Baghdad is believed to contain a significant Kurdish population, the vast majority of Kurds inhabit a region that stretches from Iraq's northern boundaries with Turkey and Iran to a line that extends roughly from the Syrian boundary west of Mosul, through Kirkuk, then down diagonally to the Iranian boundary. Along the length of this line, Kurds coexist with Turkomans and Sunni Arabs, making this one of the key ethnic flashpoints in Iraq. Added to this, most of Iraq's northern oilfields, including the Kirkuk megafield, fall within this area of ethnic heterogeneity. Sunni Arabs comprise a significant majority in Nineveh and Anbar governorates, a small majority in Salahadeen, and sizeable minorities in Diyala, Kirkuk and Baghdad. Most of the southern gover-

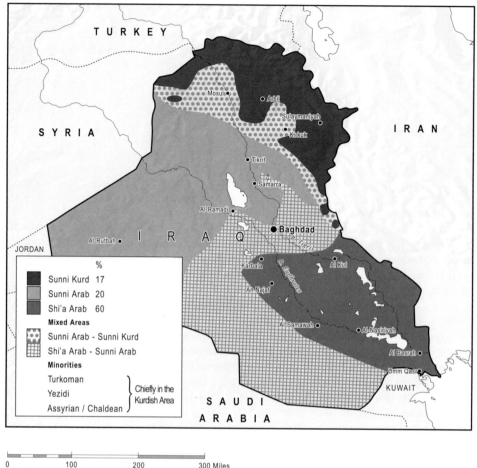

Map 81 Iraq: ethnic groups

norates are dominated numerically by Shi'a Arabs. At best these are broad generalisations, and in reality there are few parts of Iraq that are entirely "pure" in terms of ethnic identity.

The issue of ethno-sectarian identity has been a source of great controversy since the fall of the previous regime. Two broad trends have been evident. First, there has been a tendency for cities and governorates to become more homogenous in terms of communal identity. This is primarily the result of ethnic (Arab versus Kurd) and sectarian (Sunni versus Shi'a) violence in formerly mixed areas. This trend has been most pronounced in Baghdad where, in response to multiple Sunni insurgent attacks on the Shi'a population, a sustained campaign of ethnic cleansing conducted by extra-legal Shi'a militias has led to a mass Sunni Arab exodus from the capital. Similar trends have been evident in other mixed areas of Iraq, such as Nineveh and Diyala governorates.

The second trend has been the politicisation of ethno-sectarian identity. To date, Iraq's two elections have been contested by parties defined by sect or ethnicity that have earned the votes of their ethnic or sectarian cohort. Few plausible parties have emerged with genuine cross-communal appeal. As a result of this trend, the most pronounced attribute of contemporary Iraqi politics is its ethno-sectarian flavour. Efforts by Iraqi nationalist leaders such as Muqtada al-Sadr to forge broad political movements with trans-communal appeal have yet to yield tangible dividends, and nationalist forces, though probably numerous, are deeply fragmented. The success of Nuri al-Maliki's "State of Law" coalition in the 2009 governorate elections, however, suggests that a nationalist trend is solidifying around the person of the current prime minister. Al-Maliki campaigned on an overtly nationalist platform that emphasised strengthening of the central government at the expense of the regions and governorates. The national elections of January 2010 will provide a crucial test of the durability of this trend. It is clear though, that unless, and until, a credible political movement emerges with an appeal that transcends communal groups, Iraq remains in serious danger of disintergration.

Disputed territories

Iraq's disputed territories comprise a broad swathe of land stretching from Sinjar close to the Syrian boundary diagonally south-eastwards across Iraq to Badra on the Iranian boundary (Map 82). Broadly speaking, this territory is where Iraq's Kurdish and Arab populations collide. The territories also contain a significant Turkoman population, and most of the main oilfields in northern Iraq. At the heart of the territory is the hotly contested city and governorate of Kirkuk, which sits atop the north's only mega-oilfield. "Ownership" of the territory is disputed between the Kurds, who want to incorporate the territory into the autonomous Kurdish region, and most of the other political forces in Iraq, who want most of these territories (and Kirkuk in particular) to remain outside the official boundaries of Iraqi Kurdistan.

The Kurds' claim to these territories is historical (these territories are often depicted on Ottoman maps as part of Kurdistan) and geographical, but the heart of the Kurdish claim is demographic. The Kurds believe that these territories, being majority Kurdish, should be allowed to vote on whether or not to join the autonomous Kurdistan Region. However, the issue of demography is complex. Specifically, the Ba'ath Party's concerted effort to "Arabise" the region by importing Arabs into the region and expelling other ethnicities (mainly Kurds, but also Turkomans), from strategic areas such as Kirkuk, has left a legacy of demographic complexity that will be difficult to resolve. Adding to the complexity, the internal boundaries of Kirkuk were relentlessly gerrymandered by the Ba'ath regime to increase the Arab population and decrease Kurdish and Turkoman populations. Although few deny that serious population transfers took place, there is little consensus on the numbers involved, and still less on the appropriate remedy.

For the Kurds, who suffered most under Arabisation, the solution is straightforward: internal boundaries should be restored to their pre-Ba'athist status, those Arabs imported into areas like Kirkuk should return to their place of origin, expelled Kurds should return and referenda should be held to determine the future status of disputed territories. This is approximately the procedure outlined originally in Article 58 of Iraq's interim constitution and subsequently exported into Article 140 of the permanent constitution. Article 140 additionally imposed a deadline of 31 December 2007 for resolving the status of Kirkuk and other disputed territories.

The Kurds, however, have faced opposition at every stage of the process, especially as it relates to the future of Kirkuk. Kirkuk's Turkoman and Arab populations are, for the most part, adamantly opposed to holding a referendum on the future status of Kirkuk. Both groups reject Kurdish claims that hundreds of thousands of Kurds were expelled from Kirkuk and accuse the two major Kurdish parties of "Kurdifying" the city and governorate. This process, according to

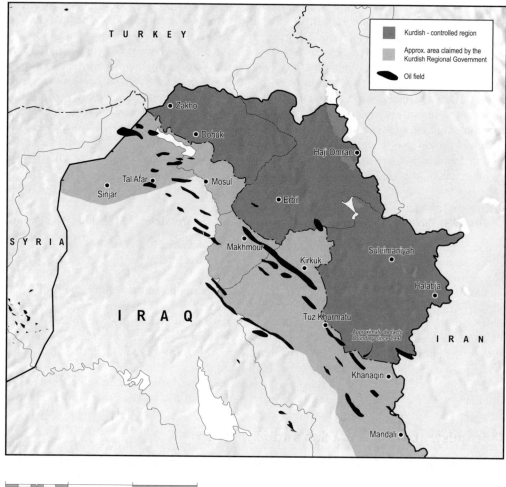

Map 82 Iraq: Kurdish area

Turkomans and Arabs, has involved moving large numbers of non-Kirkuki Kurds into Kirkuk (up to 600,000 by some estimates) in order to guarantee a positive outcome in any future referendum. Though no census has yet been held, the results of the two elections held in 2005 suggest that Kurds now comprise a small majority in Kirkuk governorate and are perhaps in a position to win a referendum on joining the Kurdistan Region. Hence, all those opposed to Kurdish ambitions have a strong interest in preventing the referendum from taking place. The complexity of the Article 140 process gave ample scope for opponents to stall for time, and ultimately, the December 2007 deadline expired without a resolution to the problem. The UN is now involved in examining potential solutions to the problem of Iraq's disputed territories, though few hold out much hope that any solution will be acceptable to all sides.

Of the various disputed territories, some are less in dispute than others. Potentially the easiest to resolve are those territories, such as the districts of Aqra and Sheikhan in Nineveh, that fall within the territory of the Kurdistan Region as governed since 1991, but lie outside the three recognised Kurdish governorates. These territories are "disputed" in the technical sense, but not in the sense of being claimed by rival parties. A similar category comprises territories, such as Makhmur, that are part of one of the three Kurdish governorates (Erbil in this case), but that fall outside the established borders of the Kurdistan Region. Here too, there is little in the way of dispute about the "Kurdishness" of the territory. More problematic are territories such as Khana-qin (Diyala), Mandali (Diyala) and Badra (Wasit) that lie outside the recognised borders of the Kurdistan Region and form part of other governorates. Sinjar, however, is probably in a cate-gory of its own in terms of the difficulties it creates. Like Khanaqin, Mandali and Badra, Sinjar is geographically part of a non-Kurdish governorate (Nineveh), and lies outside currently config-ured Kurdish borders. The incorporation of Sinjar into the Kurdistan Region would add a new layer of complexity to the situation owing to the lack of geographical contiguity between Sinjar district and the rest of the Kurdistan Region. Separating the two is the predominantly Turkoman district of Tal Afar.

However, the two most contentious disputed territories are Mosul and Kirkuk. The Kurdish claim bisects the city of Mosul along the line of the River Tigris. This makes sense demographi-cally because the east of the city is heavily populated with Kurds, but politically, the prospect of dividing a mainly Sunni Arab city is likely to prove explosive. Kirkuk, meanwhile, is the most contentious of all disputed territories on account of its oil resources, its recent history of forced expulsions and its disputed demographics. For the Kurds, the failure of an Arab-dominated Iraqi government to implement the terms of Article 140 raises important questions about the integrity of the constitution and, therefore, their future participation in the state of Iraq. For their oppo-nents, yielding oil-rich Kirkuk to the Kurds furnishes them with the resources necessary for an independent state, thereby precipitating the break-up of the Iraqi state. Thus, the stakes are high on both sides and the issue of Kirkuk's future status risks unravelling the tenuous compromise on which Iraq's nascent democracy depends for survival.

Administrative boundaries and federalism

Prior to the invasion of Iraq by US forces in 2003, the idea of federalism in Iraq had never been a serious issue. Power in Iraq's political system had been highly centralised since the creation of the state in the 1920s, and this tendency had reached it zenith under Saddam Hussein. For administrative purposes, Iraq was divided into eighteen governorates at the time of the invasion (Map 83). Though each governorate had its own separate political structure, in practice, all deci-sions were made in Baghdad and then imposed on governorates. The number of governorates and the boundaries dividing them varied considerably during Ba'athist rule. Often, governorate boundaries were manipulated with the expressed intent of altering the demographic make-up of the governorate. For example, oil-rich Kirkuk's boundaries were redrawn on numerous occa-sions to increase the size of the Arab population at the expense of Kurds and Turkomans.

Beyond naked political manipulation, the governorate boundaries inherited by post-war Iraq were largely meaningless. In the north, since 1993, the Kurds had governed a sizeable swathe of territory that encompassed three governorates almost in their entirety (Dohuk, Erbil and Sulei-maniyah) but that also included territory from other governorates (Nineveh and Diyala, for example). The so-called green line separating the Kurds from the rest of Iraq did not, therefore, coincide with existing governorate boundaries. The issue of federalism emerged in the after-math of the Iraq War as a direct consequence of Kurdish demands to preserve their autonomy in the north and to resist the re-emergence of strongly centralised power in Baghdad. Having suffered greatly under a succession of strongly centralised, Arab-dominated governments, the Kurds had a particular interest in breaking this traumatic historical pattern. Iraq's first post-war

Map 83 Governorates of Iraq

constitution, the so-called Transitional Administrative Law (TAL), called for the establishment of a decentralised federal system of government for Iraq and recognised the Kurdistan Regional Government (KRG) as the official government of the territories governed by the Kurds since 1993. Beyond this, the TAL allowed up to three governorates to join together to form larger regions, but exempted Kirkuk and Baghdad from this process.

Iraq's permanent constitution, approved by popular referendum in October 2005, envisaged a similar system for the formation of federal sub-units, but differed from the TAL in one key respect. Article 119 of the constitution allowed an unlimited number of governorates to form larger regions on the basis of popular votes in each of the governorates concerned. This opened up the possibility of a nine-governorate Shi'a region emerging in the south, comprising all governorates south of Baghdad. This provision was championed by the Supreme Council for the Islamic Revolution in Iraq (SCIRI), the most powerful political force in the Shi'a coalition, and supported by the Kurds.

The prospect of a federal system composed of two powerful regions that together contained the vast majority of Iraq's oil reserves was deeply threatening to Iraq's Sunni Arab population. The nine-governorate Shi'a super-region has also been strongly opposed by various Shi'a factions, including nationalists such as Muqtada al-Sadr. Hence, the design of Iraq's federal system remains among the most contentious of all post-war political issues. Part of the problem lies in the ambiguity of the permanent constitution with respect to the division of powers between the federal government and the regions. Article 110 of the constitution identifies certain exclusive powers that belong to the federal government (such as national defence, foreign policy and signing treaties); Article 114 outlines certain powers shared between the federal government and the regions; finally, Article 115 allocates to the regions all powers not identified as "exclusive" to the federal government under Article 114. In addition there appear to be certain powers that can be exercised by regions, but not by governorates that do not vote to form a region. Article 121, for example, gives regions power over their own internal security and accords primacy to regional law when it comes into conflict with federal law in matters outside the exclusive powers of the federal government.

Overall, the design of Iraq's federal system gives extensive potential powers to the regions. Depending on how the constitution is interpreted, these could also include significant power for regions over the management of oil and gas fields. The Kurds have already interpreted the constitution to allow for signing contracts with foreign companies to develop oilfields that lie within the Kurdistan Region. In the absence of a supreme court to adjudicate definitively on the meaning of the relevant constitutional provisions, the Kurdish interpretation remains plausible. Those who oppose the federal system as outlined in the current constitution (certain Shi'a factions and all Sunni Arab political parties) argue that it gives too much power to regions and not enough to the Iraqi government in Baghdad; those who support the constitution argue that it is precisely the concentration of power in the hands of a succession of Sunni Arab-dominated governments that has created Iraq's problems historically. Thus, the issue of federalism remains among the most divisive of all issues in post-war Iraq.

Since mid-2008, the nationalist forces have reasserted themselves under the leadership of Prime Minister Nuri al-Maliki. During 2008, al-Maliki launched a series of military assaults on rival Shi'a militias in the south and confronted the Kurds on the north under the nationalist banner of strengthening the Iraqi state. The approach paid dividends in the January 2009 governorate elections when al-Maliki's coalition took control of most governorates from Baghdad southwards, and the prime minister looks well placed to consolidate his grip on power by playing the nationalist card in the scheduled January 2010 national elections.

Afghanistan

The Afghan civil war (1978–2001)

Though somewhat arbitrary, the start of the Afghan civil war can reasonably be dated to 1978 when the communist regime in Afghanistan, headed by President Nur Muhammed Taraki, initiated a series of modernising reforms designed to "uproot feudalism" in the country. These reforms proved deeply unpopular in many parts of the country, and by the end of 1978, armed resistance was widespread. In September 1979, Deputy Prime Minister Hafizullah Amin removed Taraki from power and attempted to suppress the growing insurgency with increasingly brutal measures. As armed resistance spread from rural areas to the cities during 1979, Amin requested military assistance from the Soviet Union. Small detachments of Soviet forces began to arrive during the summer 1979, but the main Soviet force did not "invade" Afghanistan until December 1979. Although Amin had repeatedly called for Soviet intervention, the first action of incoming Soviet forces was to assault the presidential palace and kill Amin. Amin's hard-line approach to quashing the insurgency had been viewed in Moscow with increasing alarm, and he was replaced with the more moderate Babrak Karmal.

From 1980 onwards, as the Afghan armed forces proved increasingly unreliable, Soviet forces were sucked more and more into a war that they could not possibly win. The Soviets controlled the cities and attempted to control supply lines, but everything outside the cities was under the control of resistance, collectively known as the Mujahideen. The Mujahideen was a complex amalgam of Islamist groups and warlord armies that fought in a localised fashion rather than as a coherent army. Among the most prominent and effective components was a force of nearly 10,000 men based in north-eastern Afghanistan under the command of Afghan nationalist Ahmed Shah Massoud. Massoud's force was to provide the backbone of resistance to Taliban rule and the core of the Northern Alliance (NA) used by the USA during the war in 2001. Against the might of the Soviet armed forces, the Mujahideen became increasingly reliant on external support. US support for the Mujahideen actually began during the administration of President Jimmy Carter and before the Soviet invasion had taken place, but escalated rapidly under President Ronald Reagan. Orchestrated by the CIA, British intelligence and the Pakistani Inter-Services Intelligence (ISI), money and weaponry began to flow to the Mujahideen during the early to mid-1980s from a number of sources. These included the governments of the USA and the UK, but also those of many of the Gulf Arab States. While the USA generally favoured channelling support to moderate, nationalist leaders like Massoud, the lion's share of the US money and weapons were distributed to the Mujahideen by the ISI to ensure plausible deniability. This enabled the ISI to target support to certain, often extreme, Islamist factions of the Mujahideen, such as the force led by Gulbaddin Hekmatyar.

Also during the 1980s, fighters from across the Muslim world arrived in Afghanistan to join the *jihad* against communist forces, among them Osama bin Laden and a group of fanatical

Arab fighters that became the genesis of Al Qaeda (literally "the base"). Faced with escalating international support for the Mujahideen and a pointless, unwinnable war, the Soviet Union began to search for an exit strategy from about 1985 onwards. After preparing Afghan government forces to operate without Soviet assistance, the Soviet troops finally withdrew from Afghanistan in February 1989 leaving the communist government to its fate. Between 1989 and 1992, the communist regime, headed by elected president Muhammed Najibullah, succeeded in staving off defeat due mainly to the large quantity of heavy weapons left by the Soviets, and massive economic aid from the Soviet government. When the Soviet Union collapsed in 1991 and President Boris Yeltsin ordered the termination of all aid to the beleaguered Najibullah, Afghan government forces suffered massive defections, most notably by the forces of Uzbek warlord Rashid Dostum, and were soon overwhelmed as various factions of the Mujahideen descended on Kabul.

Though unified in their opposition to Soviet forces and the communist regime, the Mujahideen was hopelessly divided along multiple lines of cleavage: ethnic, religious and ideological. This became evident in 1992 when different Mujahideen factions entered Kabul from different directions and began to struggle for control of the capital. Hekmatyar's Hezbi Islami force entered Kabul first, but was subsequently driven out by other factions and resorted to shelling Kabul with artillery. The complete inability of the various Mujahideen factions to cohere together into an effective governing force meant that Afghanistan was plunged into anarchy for the next two years, which paved the way for the emergence of the disciplined and fanatical Taliban (literally "students") as Afghanistan's dominant political force.

A Sunni Islamist, predominantly Pashtun movement, the Taliban drew widespread support from the large Pashtun population of southern Afghanistan as well as, initially, the support of many Afghans who simply wanted an end to anarchy. In late 1994, the Taliban captured Kandahar and, within three months, had established control over roughly one-third of Afghanistan's provinces. In September 1996, the Taliban stormed into Kabul and seized control of all government offices within a matter of hours, driving out other Mujihadeen forces in the process. Massoud retreated to the north where he assembled a core of resistance to Taliban rule, while Dostum withdrew his forces to a swathe of territory along the Uzbek boundary. Subsequently, Dostum joined forces with Massoud to form the NA. Between 1997 and 2001, NA and Taliban forces clashed repeatedly, with neither side able to deliver the decisive blow against the other (Map 84). However, in May 1997, the Taliban assaulted Dostum's forces near the city of Mazar-e-Sharif and forced Dostum himself to seek refuge across the boundary in Uzbekistan. By September 2001, the Taliban controlled close to 95 per cent of Afghanistan's territory and the sole barrier between the Taliban and total control was the stubborn resistance of the Northern Alliance's 10,000 or so troops under the inspired leadership of Massoud.

The war in Afghanistan (2001–present)

On 9 September 2001, Ahmed Shah Massoud, the "Lion of Panjshir", was the target of a suicide attack by two assassins posing as journalists; he died a day later from his injuries. The following day, hijacked airplanes hit the twin towers of the World Trade Center in New York, and the Pentagon in Washington DC. Both attacks were perpetrated by Al Qaeda-affiliated operatives. Al Qaeda, which translates literally as "the base", began to evolve in Afghanistan after 1996 with the arrival of its spiritual inspiration and financial patron, Osama bin Laden. Initially, Al Qaeda was a relatively small group of dedicated (mainly Arab) fighters clustered around bin Laden, many of whom had participated in the *jihad* against the Soviet Union. Following the Taliban's seizure of power in 1996, the group was able to forge a working alliance with Afghan-

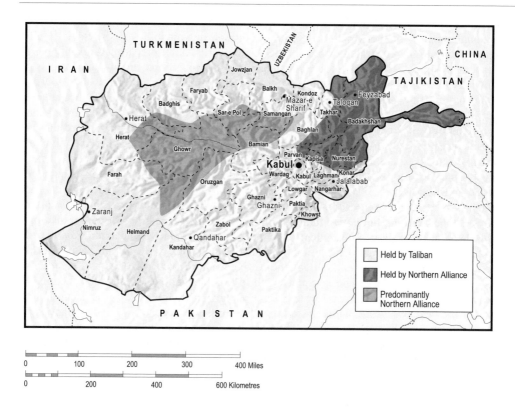

Map 84 Afghan civil war

istan's new rulers and establish a network of terrorist training camps in the country. Fighters trained in these camps were believed to be responsible for a series of terror attacks, including two mass-casualty suicide bombings at US embassies in Tanzania and Kenya in 1998. The US response, to launch a volley of cruise missiles at suspected Al Qaeda training facilities in Afghanistan, had no discernable effect on the group's operating capabilities, but did much to elevate bin Laden's profile in the Muslim world. The attacks of 11 September changed the equation significantly. Though most of the planning for the attacks was carried out in Europe, the initial US response was to identify bin Laden and Al Qaeda as the prime suspects and Afghanistan as the immediate target. On 20 September, President George W. Bush issued an ultimatum to the Taliban demanding that they shut down all terrorist training facilities in Afghanistan and turn over Al Qaeda leaders to US authorities. Following the Taliban's refusal to comply, a US–British bombing campaign began on 7 October.

Anxious to avoid the fate of Soviet forces during the 1980s, the USA relied mostly on airpower to fight the first stages of the war, supported on the ground by teams of special forces and members of the CIA's Special Activities Division. The strategy was to avoid a large-scale commitment of ground troops and to rely instead on forces of the NA and other Afghan tribal militias to do the foot soldiering. After a month of extremely heavy bombardment of Taliban positions around the northern city of Mazar-e-Sharif, the NA entered the city on 9 November meeting only token resistance from surviving Taliban forces. The NA advanced rapidly through the northern provinces and arrived in Kabul on 13 November to find that the Taliban had melted away. With the capture of Kabul, the Taliban's hold on power effectively disintegrated. The last holdout was the Taliban's city of origin, Kandahar, which fell to coalition forces during the first week of December.

Militarily, this was a heavy defeat for the Taliban regime that had been devastated by the sustained exercise of massive US air power. It was unclear, however, whether the bulk of Taliban forces had been killed, had defected or had carried out a strategic retreat across the boundary with Pakistan. Indeed, one of the major problems with the US campaign was the failure to seal off the Afghan–Pakistan boundary. Lacking troops on the ground to perform the job itself, the USA had relied on the Pakistani military to close off the boundary. This turned out to be a serious miscalculation. It later transpired that most of the Taliban's senior leadership, including Mullah Omar, had escaped across the boundary into Pakistan's mainly Pashtun tribal areas. The USA also missed an opportunity to eliminate Al Qaeda and perhaps even bin Laden himself in the mountains of Tora Bora. As a force of approximately 200 fighters covered their retreat, bin Laden led the bulk of Al Qaeda's forces to safe refuge across the porous Pakistan boundary. Subsequently, the lawless tribal region of Pakistan became the strategic base from which the Taliban and Al Qaeda mounted their insurgency. Consequently, what had seemed like a clear and decisive US victory against the Taliban and Al Qaeda was transformed into a similar sort of war to that waged by the Soviets in the 1980s.

During the time it took the Taliban to rebuild and reorganise their forces, Afghanistan achieved a form of political stability. A council of prominent tribal and factional leaders (*Loya jirga*) was convened in 2002 to select Hamed Karzai as interim president, and a subsequent constitutional *Loya jirga* drew up Afghanistan's first post-war democratic constitution. Presidential elections, won by Karzai, followed in 2004, and parliamentary elections in 2005. The USA, meanwhile, canvassed the international community for the vast quantities of economic aid that would be needed to rebuild Afghanistan's shattered infrastructure after decades of sustained warfare.

After the US focus shifted to the war in Iraq, however, the political and security situation in Afghanistan began to deteriorate rapidly. From summer 2003 onwards, the Taliban began to launch increasingly sophisticated attacks from their bases in Pakistan. Pakistan's armed forces were either unable or unwilling to mount a full-scale offensive into the traditionally ungovernable tribal areas, and Pakistan's government was forced to make deals with prominent tribal leaders rather than confront them directly. As a result, Taliban fighters could cross the boundary with virtual impunity, carry out attacks and then retreat back into the relative safety of Pakistan. With the Pakistan government unable to stem the flow of fighters across the boundary, the USA was forced to rely increasingly on missile strikes into Pakistan, which were sometimes militarily effective, but tended to generate huge resentment in the tribal areas and throughout Pakistan.

The pattern established by the end of 2005 was that the Taliban would generally suffer heavy defeats when they confronted US forces directly on the battlefield and in large numbers, but operating in small guerrilla units, they could inflict serious damage on the credibility of US and Afghan government claims that Afghanistan was secure. Exploiting their natural support base among the Pashtun population, the Taliban steadily gained ground during 2006–8. By the end of 2008, the Taliban was estimated to constitute a permanent presence in over half of the territory of Afghanistan (Map 85) and had begun the process of sealing Kabul off from its supply routes.

At the beginning of 2006, the relatively small US force operating in the dangerous southern provinces was replaced by NATO troops, mainly drawn from Britain, Canada and the Netherlands. The original intention was to form provincial reconstruction teams to spearhead the rebuilding of southern Afghanistan in an effort to win hearts and minds. However, operating in difficult terrain and under constant attack from insurgent forces, NATO troops have been unable to make significant progress. Efforts to win hearts and minds have also been seriously under-

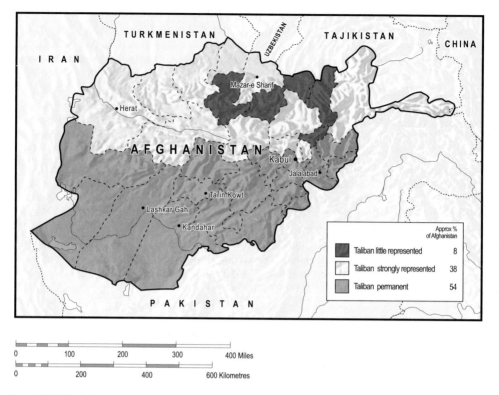

Map 85 Taliban distribution (2008)

mined by the periodic accidental bombing of civilians during US air strikes. By the start of 2009, the situation in Afghanistan was in a downward spiral. The Taliban was in firm control of several southern provinces, and attacking coalition supply lines in Pakistan (through which three-quarters of all coalition supplies come) with increasing effectiveness. Afghanistan was, by any measure, a failed state with a government that was unable to project power beyond Kabul and increasingly unable to protect itself within Kabul. The newly elected US President Obama pledged to reverse the tide by sending some 40,000 extra US troops to Afghanistan. It remains to be seen whether this force will be sufficient to make a decisive difference, or whether it is just paving the way for the sort of unwinnable war inflicted on the Soviet Union twenty years previously.

The Afghan drug trade

Alongside the ongoing, and increasingly violent, Taliban insurgency, one of the major barriers to reconstruction and the establishment of security in Afghanistan is the burgeoning trade in drugs. The cultivation of opium poppies in Afghanistan is certainly not new; what has changed over recent decades is the scale of the problem. By 2008, Afghanistan was estimated to be responsible for over 90 per cent of the global production of opium. Moreover, the trend over recent years has been towards processing the opium into higher value opiates, heroin and morphine, prior to export. Hence, Afghanistan is now Europe's major supplier of heroin and one of the world's most significant producers of illicit morphine. As with all narco-states, the illegal drug industry in Afghanistan both benefits from, and contributes to, the wholesale corruption of governing institutions, which in turn greatly impedes the possibility of a stable, reliable democratic order emerging in the short term. The proceeds of the illicit drug trade also fund the activities of warlord militias and Taliban insurgents. Reportedly, the Taliban imposes a "tax" of 10 per cent on the drug trade in the areas it controls in the south of the country. The drug trade is, therefore, a major threat to Afghanistan's political stability. However, the usual remedies, such as stricter enforcement and crop spraying, may arguably create more problems than they resolve.

The spectacular growth of the drugs industry in Afghanistan can be traced back to the 1980s and the years of the Soviet occupation. The vast profits available from the cultivation and sale of illicit drugs made them an appealing source of income to warlords and regional commanders of the Mujahideen to fund the struggle against the Soviet Union. As the central government lost control of large swathes of territory, the areas under cultivation expanded rapidly. The CIA agents in Afghanistan who orchestrated the arming and funding of the Mujahideen turned a blind eye to the drug trade and were even accused (by the Soviets) of active participation in the smuggling of heroin out of Afghanistan to European markets. The situation degenerated further once funding from the Soviet Union and the USA dried up in the early 1990s. As the Mujahideen disintegrated into multiple warring factions, warlords sought to maximise the profits from cultivating and exporting opiates in order to fund and equip their armies, and the scale of Afghanistan's drug problem escalated.

The Taliban's relationship with the drug trade is ambiguous. During its rise to power, it appears that the Taliban gained support in Pashtun areas by protecting regional drug lords and used the proceeds of drug sales to fund their various military operations. Moreover, as the Taliban extended its grip over the country, the area of land that went towards opium poppy cultivation expanded annually. In 1999, for example, the number of hectares of land given over to poppy cultivation was higher than it had been at any point in Afghanistan's history. However, the Taliban was also directly responsible for the virtual eradication of poppy cultivation during 2001. Declaring poppy cultivation to be "un-Islamic" Mullah Mohammed Omar launched a spectacularly successful counter-narcotics campaign that reduced production of opium to virtually zero in 2001.

Subsequent to the US attack on Afghanistan and the removal from power of the Taliban, the production and sale of opiates in Afghanistan has grown to unprecedented levels. In 2002, for example, poppy cultivation reached virtually the level of 1999, but then doubled between 2002 and 2007. In the process, Afghanistan eclipsed Myanmar as the world's major producer of opiates and now supplies the vast majority of the world's illicit opium, morphine and heroin. There is a stark regional disparity in terms of production. Overwhelmingly, the southern part of the country, and especially Helmand Province, dominates supply (Map 86). By 2008, the south was responsible for 69 per cent of poppy cultivation, followed by the western region (15 per cent) and the east (11 per cent). In general, there is a strong correlation between the absence of security and the growth of cultivation. Those areas that are the least secure for coalition forces are the major opium-producing regions for fairly obvious reasons. However, while it is clear that the Taliban is drawing funding from the proceeds of the drug trade and support from disaffected farmers whose poppy crop has been the target of coalition eradication efforts, it is far from clear that the various members of the coalition are fully committed to eradicating drugs from Afghanistan.

Much of the drug trade in "secure" parts of Afghanistan is controlled by warlords that are allied with the coalition. Moreover, in the safest part of Afghanistan, Kabul, poppy cultivation has actually increased over recent years, suggesting that members of the Afghan government are

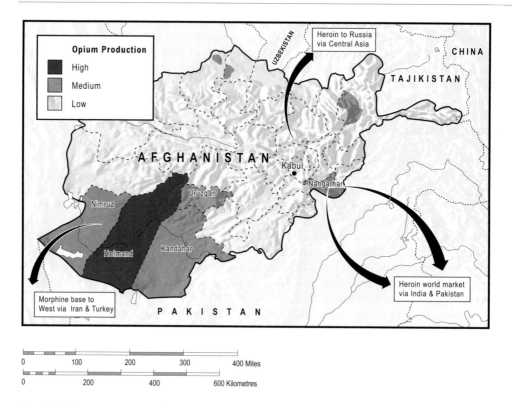

Map 86 Afghanistan: opium cultivation

deeply involved in the drug trade, and that coalition forces are willing to turn a blind eye. The deeper problem for Afghanistan is that the drug trade now provides somewhere between 40 and 50 per cent of the country's annual GDP. With these sorts of profits available from the drug trade, it is scarcely surprising that an estimated 10 per cent of Afghanis are employed at some point in the drug supply chain, or that drug money is having a predictably corrupting influence on the forces of law and order and the other institutions of government. It is not possible to eradicate poppy cultivation until the Taliban has been driven out of the areas it currently controls; but it is not possible to drive the Taliban out until it can be cut off from the source of its power, which is the drug trade. Even if this conundrum could be resolved, eliminating the drug trade would still alienate many powerful political interests and would leave millions of Afghanis destitute.

In the meantime, opiates continue to flow out of Afghanistan at an alarming (and increasing) rate with pernicious results for all surrounding countries. As indicated on Map 86, there are three main drug trafficking routes out of Afghanistan, as well as a host of smaller points of exit. The UN has estimated that between 50 and 60 per cent of Afghan opiates are moved across Afghanistan's northern boundaries with the former Soviet Central Asian Republics of Tajikistan and Uzbekistan. The product is usually refined into heroin in Afghanistan, then transported mainly through Tajikistan and on to Russia and the West. A second series of trafficking routes connect various opiate-producing Afghan provinces with Pakistan, principally the North West Frontier Province and Baluchistan. Much of the product is consumed in Pakistan, but some continues further south into India, and a portion traverses the Pakistan–Iran boundary. It is then refined into heroin in Iran, or shipped on to Turkey for refinement and eventual transport to Western Europe. The third major route, and the one through which most of Western Europe's heroin travels, is directly out across the boundary between southern Afghanistan and Iran in the form of morphine base. From there, it moves to refineries in Iran or on to Turkey for refinement. The clearing house for much of Afghanistan's opiate product appears to be Baluchistan, an ethnic region that cuts across the boundaries of Pakistan, Afghanistan and Iran and is, therefore, very difficult to police effectively.

Among the detrimental effects of the drug trade is the addiction rates of populations through which the drugs are trafficked. Iran, Afghanistan and Tajikistan have among the world's highest rates of opium and heroin addiction. Yet, stamping out the drug trade, or preventing the flow of drugs out of Afghanistan and into neighbouring countries is all but impossible. In Iran alone, there are more than 30,000 law enforcement personnel manning boundaries with Pakistan and Afghanistan in an effort to stem the tide; but the boundaries themselves are difficult to police and the quantity of drugs involved is so huge that effective interdiction is impossible. Elsewhere, such as Tajikistan and Russia, corruption among law enforcement officials is endemic (largely a product of the drug trade itself) so trafficking routes can be maintained by bribery. Viewed realistically, Afghanistan is a failed narco-state, similar in kind to Colombia during the 1980s, but much worse off due to poverty rates and pervasive insecurity.

Section F

Further reading

General reading

Agnew, C. and Anderson, E. (1992) *Water Resources in the Arid Realm*, London: Routledge.

Ali, T. (2002) *The Clash of Fundamentalisms: Crusades, Jihads and Modernity*, London: Verso.

Allan, T. (2001) *The Middle East Water Question: Hydropolitics and the Global Economy*, London: I.B. Tauris.

Anderson, E.W. (2000) *Global Political Flashpoints: An Atlas of Conflict*, London: The Stationery Office.

Anderson, E.W. (2000) *The Middle East: Geography and Geopolitics,* London: Routledge.

Anderson, E.W. (2003) *International Boundaries: A Geopolitical Atlas*, London: The Stationery Office.

Beaumont, P., Blake, G.H. and Wagstaff, J.M. (1988) *The Middle East: The Geographical Study*, London: David Fulton.

Bennis, P. (2003) *Before and After: US Foreign Policy and the War on Terrorism*, Moreton-in-Marsh: Arris Books.

Bhattacharyya, G. (2005) *Traffick: The Illicit Movement of People and Things*, London: Pluto Press.

British Petroleum (2008) *BP Statistical Review of World Energy, 2008,* London: British Petroleum.

Central Intelligence Agency (2008) *The World Factbook*, Washington, DC: CIA.

Clancy-Smith, J. (2201) *North Africa, Islam and the Mediterranean World*, London: Frank Cass.

Cohen, R. (ed.) (1995) *Cambridge Survey of World Migration*, Cambridge: Cambridge University Press.

Cooke, R., Warren, A. and Goudie, A. (1993) *Desert Geomorphology,* London: UCL Press.

Cordesman, A.H. (1999) *Transnational Threats from the Middle East: Crying Wolf or Crying Havoc,* Carlisle, PA: Strategic Studies Institute US Army War College.

Europa Regional Surveys of the World (2006) *Africa South of the Sahara 2007*, London: Routledge.

Europa Regional Surveys of the World (2006) *Eastern Europe, Russia and Central Asia 2007*, London: Routledge.

Europa Regional Surveys of the World (2006) *The Middle East and North Africa 2007*, London: Routledge.

Europa Regional Surveys of the World (2006) *South Asia 2007*, London: Routledge.

International Institute for Strategic Studies (2007) *IISS Strategic Survey, 2007: The Annual Review of World Affairs*, London: Routledge.

International Institute for Strategic Studies (2008) *The Military Balance, 2008*, London: Routledge.

Israeli, R. (2003) *War, Peace and Terror in the Middle East*, London: Frank Cass.

Karsh, E. (2003) *Rethinking the Middle East*, London: Frank Cass.

Keay, J. (2003) *Sowing the Wind: The Seeds of Conflict in the Middle East*, London: John Murray.

Lewis, B. (1997) *Middle East*, London: Phoenix.

Lewis, B. (2002) *What Went Wrong? The Clash between Islam and Modernity in the Middle East,* London: Weidenfeld and Nicolson.

Lewis, B. (2004) *The Crisis of Islam: Holy War and Unholy Terror*, London: Weidenfeld and Nicolson.

Miles, H. (2005) *Al Jazeera: How Arab TV News Challenged the World*, London: Abacus.

Milton-Edwards, B. and Hinchcliffe, P. (2001) *Conflicts in the Middle East since 1945*, London: Routledge.

Owen, R. (2004) *State, Power and Politics in the Making of the Middle East*, London: Routledge.

Sluglett, P. and Farouk Sluglett, M. (eds) (1996) *The Times Guide to the Middle East*, London: Times Books.

Smith, D. (2006) *The State of the Middle East: An Atlas of Conflict and Resolution*, London: Earthscan.

Specific reading

Abdul-Jabar, F. (ed.) (2006) *Ayatollahs, Sufis and Ideologues: State Religion and Social Movements in Iraq*, London: Saqi Books.

Abdul-Jabar, F. (ed.) (2006) *The Shi'ite Movement in Iraq*, London: Saqi Books.

Aghrout, A. and Bougherira, R. (eds) (2004) *Algeria in Transition: Reforms and Development Prospects*, London: Routledge.

Aissaoui, A. (2001) *Algeria – The Political Economy of Oil and Gas*, Oxford: Oxford University Press.

Al Abed, I. and Hellyer, P. (eds) (2001) *The United Arab Emirates*, London: Trident Press.

Al Dekhauel, A. (2000) *Kuwait: Oil, State and Political Legitimation*, London: Ithaca Press.

Alizadeh, P. (ed.) (2001) *The Economy of Iran: The Dilemmas of an Islamic State*, London: I.B. Tauris.

Allen, C.H. and Rigsbee, W.L. (2000) *Oman under Qaboos: From Coup to Constitution, 1970–1976*, London: Frank Cass.

Al-Mallakh, R. (1985) *Qatar, Energy and Development*, London: Croom Helm.

Al-Musawi, M.J. (2006) *Reading Iraq: Culture and Power in Conflict*, London: I.B. Tauris.

Al-Rasheed, M. (2002) *A History of Saudi Arabia*, Cambridge: Cambridge University Press.

Al Sa'ud, F. (2004) *Iran, Saudi Arabia and the Gulf*, London: I.B. Tauris.

Altunisik, M.B. and Kavli, O.T. (2004) *Turkey: Themes and Challenges*, London: Routledge.

Anderson, L. and Stansfield, G. (2004) *The Future of Iraq: Dictatorship, Democracy or Division?* New York: Palgrave Macmillan.

Arburish, S. (2000) *Saddam Hussein: The Politics of Revenge*, London: Bloomsbury.

Arnove, A. (ed.) (2000) *Iraq under Siege: The Deadly Impact of Sanctions and War*, London: Pluto Press.

Aruru, N.H. (ed.) (2001) *Palestinian Refugees: The Right of Return*, London: Pluto Press.

Aydin, Z. (2005) *The Political Economy of Turkey*, London: Pluto Press.

Barakat, S. (2005) *After the Conflict: Reconstruction and Redevelopment in the Aftermath of War*, London: I.B. Tauris.

Barari, H.A. (2004) *Israeli Politics and the Middle East Peace Process, 1988–2002*, London: RoutledgeCurzon.

Bianchi, R.R. (2004) *Guests of God: Pilgrimage and Politics in the Islamic World*, Oxford: Oxford University Press.

Bishara, M. (2001) *Palestine–Israel: Peace or Apartheid. Prospects for Resolving the Conflict*, London: Zed Books.

Bligh, A. (2002) *The Political Legacy of King Hussein*, Brighton: Sussex Academic Press.

Blix, H. (2004) *Disarming Iraq: The Search for Weapons of Mass Destruction*, London: Bloomsbury.

Bonner, M., Reif, M. and Tessler, M. (eds) (2005) *Islam, Democracy and the State in Algeria: Lessons for the Western Mediterranean and Beyond*, London: Routledge.

Bonora, C. (2000) *France and the Algerian Conflict*, Aldershot: Ashgate Publishing.

Bouillon, M. (2004) *The Peace Business: Money and Power in the Palestine–Israel Conflict*, London: I.B. Tauris.

Bowen, J. (2003) *Six Days: How the 1967 War Shaped the Middle East*, London: Simon and Schuster.

Bowen, W.Q. (2006) *Libya and Nuclear Proliferation: Stepping Back from the Brink*, London: Routledge.

Brandell, I. (ed.) (2006) *State Frontiers: Borders and Boundaries in the Middle East*, London: I.B. Tauris.

Bregman, A. (2003) *A History of Israel*, New York: Palgrave Macmillan.

Butler, V., Carney, T. and Freeman, M. (2005) *Sudan: The Land and the People,* London: Thames & Hudson.

Byrd, W.A. (2005) *Afghanistan: State Building, Sustaining Growth and Reducing Poverty*, Washington, DC: World Bank.

Calderini, S., Cortese, D. and Webb, J.L.A. (1992) *Mauritania,* Oxford: ABC Clio.

Carkoglu, A. and Rubin, B.M. (eds) (2004) *Greek–Turkish Relations in an Era of Détente*, London: Frank Cass.

Carkoglu, A. and Rubin, B.M. (eds) (2005) *Religion and Politics in Turkey*, Abingdon: Routledge.

Carnahan, M., Manning, N., Bontjer, R. and Guimbert, S. (eds) (2004) *Reforming Fiscal and Economic Management in Afghanistan*, Washington, DC: World Bank.

Cattan, H. (2000) *The Palestine Question,* London: Saqi Books.

Champion, D. (2003) *The Paradoxical Kingdom: Saudi Arabia and the Momentum of Reform,* London: C. Hurst.

Chehab, Z. (2005) *Iraq Ablaze: Inside the Insurgency*, London: I.B. Tauris.

Christou, G. (2004) *The European Union and Enlargement:The Case of Cyprus*, New York: Palgrave.

Cockburn, P. (2006) *The Occupation: War and Resistance in Iraq*, London: Verso.

Cohen, S. and Jaidi, L. (2006) *Morocco,* Abingdon: Routledge.

Cohn-Sherbok, D. and El-Alami, D. (2001) *The Palestine–Israel Conflict: A Beginner's Guide,* Oxford: Oneworld Publications.

Colburn, M. (2004) *The Republic of Yemen: Development Challenges in the 21st Century,* London: Catholic Institute for International Relations.

Danchev, A. and Macmillan, J. (eds) (2004) *The Iraq War and Democratic Politics*, London: Routledge.

Dib, K. (2004) *Warlords and Merchants: The Lebanese Business and Political Establishments,* London: Ithaca Press.

Dolphin, R. (2006) *The West Bank Wall: Unmaking Palestine*, London: Pluto Press.

Ebadi, S. (2006) *Iran Awakening: A Memoir of Revolution and Hope*, London: Rider.

Fahmy, N.S. (2002) *The Politics of Egypt: State-Society Relationship*, London: RoutledgeCurzon.

Fandy, M. (2003) *Kuwait and a New Concept of International Politics*, Basingstoke: Palgrave Macmillan.

Fisk, R. (2001) *Pity the Nation: Lebanon at War,* Oxford: Oxford Paperbacks.

Flint, J. and de Waal, A. (2005) *Darfur: A Short History of a Long War,* London: Zed Books.

Fraser, T.G. (2004) *The Arab–Israeli Conflict*, London: Palgrave Macmillan.

Furtig, H. (2000) *Iran's Rivalry with Saudi Arabia Between the Gulf Wars,* Reading: Ithaca Press.

Gabriel, W. (2003) *Islam, Sectarianism and Politics in Sudan since Mahdiyya*, London: C. Hurst.

George, A. (2003) *Syria: Neither Bread nor Freedom*, London: Zed Books.

George, A. (2005) *Jordan: Living in the Crossfire*, London: Zed Books.

Ghanem, A. (2002) *The Palestinian Regime: A "Partial Democracy"*, Brighton: Sussex Academic Press.

Ghubash, H. (2004) *Oman: A Millennial Islamic Democracy*, London: Saqi Books.

Gold, P. (2000) *Europe or Africa? A Contemporary Study of the Spanish North African Enclaves of Ceuta and Melilla*, Liverpool: Liverpool University Press.

Goodarzi, J. (2006) *Syria and Iran: Diplomatic Alliance and Power Politics in the Middle East,* London: I.B. Tauris.

Hannay, D. (2005) *Cyprus: The Search for a Solution,* London: I.B. Tauris.

Harik, J.P. (2004) *Hezbollah: The Changing Face of Terrorism*, London: I.B. Tauris.

Hiro, D. (2005) *Secrets and Lies: The Planning, Conduct and Aftermath of Blair and Bush's War*, London: Politico's Publishing.

Hourani, A., Khoury, P. and Wilson, M.C. (eds) (2004) *The Modern Middle East*, London: I.B. Tauris.

Houston, C. (2001) *Islam, Kurds and the Turkish Nation State*, Oxford: Berg.

Howe, M. (2005) *Morocco: The Islamist Awakening and Other Challenges*, New York: Oxford University Press.

Ikram, K. (2005) *Egyptian Economy: Performance, Policies and Issues*, London: Routledge.

Iskandar, M. (2006) *Rafiq Hariri and the Fate of Lebanon,* London: Saqi Books.

Jenkins, G. (2001) *Context and Circumstance: The Turkish Military and Politics*, Oxford: Oxford University Press.

Jenkins, G. (2006) *Political Islam in Turkey,* New York: Palgrave Macmillan.

Johnson, M. (2001) *All Honourable Men: The Social Origins of War in Lebanon,* London: I.B. Tauris.

Joseph, J.S. (2006) *Turkey and the European Union*, New York: Palgrave Macmillan.

Kayal, A.D. (2002) *The Control of Oil: East–West Rivalry in the Persian Gulf*, London: Kegan Paul.

Keegan, J. (2004) *The Iraq War*, London: Hutchinson.

Khan, M. (ed.) (2004) *State Formation in Palestine: Establishing Good Governance and Democracy through Social Transformation*, London: RoutledgeCurzon.

Kienle, E. (2001) *A Grand Delusion: Democracy and Economic Reform in Egypt*, London: I.B. Tauris.

Kurz, A.N. (2005) *Fatah and the Politics of Violence: The Institutionalisation of a Popular Struggle,* Brighton: Sussex Academic Press.

Lahn, G. (2004) *Democratic Transition in Bahrain: A Model for the Arab World*, London: Gulf Centre for Strategic Studies.

Lobmeyer, H.G. (2004) *Opposition and Resistance in Syria*, London: I.B. Tauris.

McGilvary, M. (2002) *The Dawn of a New Era in Syria,* Reading: Garnet Publishing.

Mackey, S. (2002) *The Saudis: Inside the Desert Kingdom,* London: W.W. Norton.

Makdisi, S. (2004) *Lessons of Lebanon: The Economics of War and Development*, London: I.B. Tauris.

Manea, E. (2005) *Regional Politics in the Gulf: Saudi Arabia, Oman and Yemen*, London: Saqi Books.

Masalha, N. (2000) *Imperial Israel and the Palestinians: The Politics of Expansion*, London: Pluto Press.

Miles, H. (2005) *Al Jazeera: How Arab TV News Challenged the World*, London: Abacus.

Milton-Edwards, B. and Hinchcliffe, P. (2001) *Jordan: A Hashemite Legacy*, London: Routledge.

Mohammadi, A. (2006) *Iran Encountering Globalisation: Problems and Prospects*, London: Routledge.

Mohammadi, A. and Ehteshami, A. (eds) (2000) *Iran and Eurasia*, London: Ithaca Press.

Moller, B. (ed.) (2001) *Oil and Water: Cooperative Security in the Persian Gulf*, London: I.B. Tauris.

Moore, P. (2001) *Bahrain: A New Era*, London: Euromoney Books.

Morkot, R. (2005) *Egyptians: An Introduction*, London: Routledge.

Moustakis, F. (2003) *The Greek–Turkish Relationship and NATO*, London: Frank Cass.

Murphy, E.C. (2000) *Economic and Political Change in Tunisia: From Bourguiba to Ben Ali,* London: St Martin's Press.

Nachmani, A. (2003) *Turkey: Facing a New Millennium: Coping with Intertwined Conflicts*, Manchester: Manchester University Press.

Nasr, V. and Gheissari, A. (2006) *Democracy in Iran: History and the Quest for Liberty*, Oxford: Oxford University Press.

Naumkin, V. (2004) *Red Wolves of Yemen: The Struggle for Independence*, Cambridge: Oleander Press.

Niblock, T. (ed.) (2006) *Saudi Arabia: Power, Legitimacy and Survival,* London: Routledge.

Obeidi, A. (2001) *Political Culture in Libya*, Richmond: Curzon.

Onis, Z. and Rubin, B.M. (eds) (2003) *The Turkish Economy in Crisis*, London: Frank Cass.

Oren, M.B. (2002) *Six Days of War: June 1967 and the Making of the Modern Middle East*, Oxford: Oxford University Press.

Ould-May, M. (1996) *Global Restructuring and Peripheral States: The Carrot and Stick in Mauritania,* Lanham, MD: Littlefield Adams.

Owtram, F. (2004) *A Modern History of Oman: Formation of the State since 1920,* London: I.B. Tauris.

Pappe, I. (2004) *A History of Modern Palestine: One Land, Two Peoples,* Cambridge: Cambridge University Press.

Parra, F. (2003) *Oil Politics: A Modern History of Petroleum*, London: I.B. Tauris.

Perkins, K.J. (2004) *A History of Modern Tunisia*, Cambridge: Cambridge University Press.

Potter, L.G. and Sick, G.E. (eds) (2006) *Iran, Iraq and the Legacies of War*, London: Palgrave Macmillan.

Prunier, G. (2005) *Darfur: The Ambiguous Genocide*, London: Hurst.

Reilly, H. (2005) *Seeking Sanctuary: Journeys to Sudan*, Bridgnorth: Eye Books.

Robins, P. (2004) *A History of Jordan*, Cambridge: Cambridge University Press.

Rothstein, R.L., Ma'oz, M. and Shikaki, K. (eds) (2002) *The Israeli–Palestinian Peace Process: Oslo and the Lessons of Failure*, Brighton: Sussex Academic Press.

Roy, O. (2005) *Turkey Today: A European Nation?* London: Anthem Press.

Salam, N.A. (ed.) (2005) *Options for Lebanon*, London: I.B. Tauris.

Salhiz, S. and Netton, I.R. (2005) *The Arab Diaspora*, London: Routledge.

Searight, S. (2002) *Yemen: Land and People*, London: Pallas Athene.

Selby, J. (2003) *Water, Power and Politics in the Middle East: The Other Palestinian–Israeli Conflict*, London: I.B. Tauris.

Shahak, I. and Mezvinsky, N. (2004) *Jewish Fundamentalism in Israel*, London: Pluto Press.

Shulze, R. (2002) *A Modern History of the Islamic World*, London: I.B. Tauris.

Sidahmed, A.S. and Sidahmed, A. (2004) *Sudan,* London: Routledge.

Simons, G.L. and Benn, T. (2004*) Libya and the West: From Independence to Lockerbie,* London: I.B. Tauris.

Slot, B.J. (ed.) (2003) *Kuwait: The Growth of a Historic Identity*, London: Arabian Publishing.

Soul, K.H. (2004) *The Arab World: An Illustrated History*, New York: Hippocrene Books.

Tal, N. (2005) *Radical Islam in Egypt and Jordan*, Brighton: Sussex Academic Press.

Walker, J. (2004) *Aden Insurgency: The Savage War in South Arabia 1962–67,* Staplehurst: Spellmount Publishers.

Wasserstein, B. (2004) *Israel and Palestine: Why They Fight and Can They Stop?* London: Profile Books.

Wilson, R., al-Salamah, A., Malik, M. and al-Rajhi, A. (2003) *Economic Development in Saudi Arabia,* London: RoutledgeCurzon.

Yamani, M. (2004) *Cradle of Islam: The Hijaz and the Quest for Arabian Identity*, London: I.B. Tauris.

Yavuz, M.H. (2003) *Islamic Political Identity in Turkey*, Oxford: Oxford University Press.

Yildiz, K. (2005) *The Kurds in Turkey: EU Accession and Human Rights*, London: Pluto Press.

Zisser, E. (2000) *Lebanon: The Challenge of Independence*, London: I.B. Tauris.

Zisser, E. (2000) *Asad's Legacy: Syria in Transition*, London: C. Hurst.

Index

Bold type denotes the pages on which the main discussion of any particular topic is to be found. For the states of the Middle East and the Rimland regions, the material in Section D is not indexed separately but is subsumed in a main discussion reference (in bold) for each state.